CASSIOPEIA'S QU

An Homage to the Strong Wome

Li

This collection of 10 short work

young female protagonist in each story through some of the events in the evolution of human culture over 100,000 years.

Deep in the mists of time, perhaps a million or so years ago, our proto-human species was forced to descend from the warm, enfolding and close embrace of the treetops, then disappearing with the climate change of those days. Once on the ground we began to wonder, as homo sapiens, at the stars in the night sky and what might lie beyond the far horizons of the open grassy savanna.

By these 10 stories each of our young protagonists will seek to inform and illuminate our humanity.

Jerry Smetzer

Writer, Painter, Science Enthusiast

Copyrights

This is a collection of short works of speculative fiction. There are many fictional characters in each of 10 stories who are products of my imagination. They bear no relationship to any real person either alive or dead. The fictional characters interact among themselves and among other named characters who are said to have existed in history or who have been referenced in the record of historic facts, figures and events.

I am grateful to Kaci Hamilton, formerly of Juneau, Alaska for permission to use her image for the face of the figure of Cassiopeia on the cover. I am also grateful to Annie Kincheloe for her help designing the cover and coordinating the printing with Alaska Litho of Juneau. The background image on the cover is an iStock image purchased and authorized for use by Getty Images. Image owner is: Michael Ver Sprill.

ISBN: 978-1-7326656-4-4

Names: Smetzer, Jerry, author

Title: Cassiopeia's Quest – Revelation:

a collection of 10 short works of speculative/historic fiction

by Jerry Smetzer

www.jerrysmetzer.net | jsmetzer@gmail.com

First edition

Dedication

To Clara

I hope you enjoy this book!

For all the children and grandchildren

Table Of Contents

Selected Paragraphs from Stories in the collection.

From: *EMPIRE...*

Page 84 Third paragraph from the bottom.

Alexander walks back toward the group, again lightly swinging and twirling the blued steel while he examines the length of it. He reads the engraving aloud: "Plato is dear to me, but dearer still is truth." He gives a smart grin. As he returns the sword to its sheath, he says, "A fine statement by my teaching master, Aristotle; a fine statement for a fine weapon.

Thank you, Tabu. I hope you and your master will agree to join us on our campaign to the east."

As Alexander turns back to face the approaching Kallias, now armed and armored, Tabu speaks. "Thank you, my Pharaoh. It is my great honor to serve you."

Alexander's first thrust to the chest is easily parried by Kallias, who then moves to block Alexander's next killing thrust low toward his belly. My breath is drained away. I sense my open mouth and think to myself that I must close it.

These two men are closely matched and well aware of the many practiced and usually successful thrusts and parries that they have had to both execute and also protect against. After a few minutes of intense swordplay, Alexander suddenly strikes, hard, at Kallias' kopis. The bronze blade breaks under the power of Alexander's stroke. The broken blade falls to the ground and lays there.

Alexander's theatrical nature breaks through as he thrusts his sword against Kallias' now undefended neck. He stops within a finger's width of raw flesh, and then lays the flat of the blade against the skin of Kallias' neck.

From: *FAITH*...

Page 99. First paragraph on the first page.

Ziva suddenly calls out, "Gila! This way! Quickly!"

The younger Gila runs to her sister's side to see what she is yelling about. When she sees, her eyes light up. Around the corner, in one of the open shops in the bustling market of lower Jerusalem, a young, bearded man in a light gray robe is talking quietly to a crowd of people. The small audience has turned its ear to the soft words being spoken to them. Gila preacher from Galilee known as Jesus of Nazareth. Her father has worked as a craftsman in the Galilee and speaks often of the time he went to temple and Jesus was there. Now, she and her sister have found him, by luck, in the Passover marketplace.

From: *REBIRTH*...

Page 165. First two paragraphs.

"She is an orphan, a girl of the streets living in a poor section of Florence, near the industrial docks along the Arno. She does not know her parents, nor can she say with any certainty the last time she felt close to a grown man or woman, other than the young woman, a prostitute catering to the upper classes, who took her in when she was barely more than a toddler.

"Their paths had crossed early one morning on Florence's back streets, as the taverns closed and before the shop stalls opened. The orphan had suddenly appeared beside the prostitute and taken her hand. The orphan looked up into the eyes of the prostitute who was, at first, not sure what to do with this small child. As she walked to her small but clean room after a night of entertaining the guests of her most important client, she was not at all sure she had the energy to deal with this."

From "Now I Am Become Death"...

Page 237. Third paragraph up from bottom

"Feynman paused. "I've read your papers on Maxwell's work in optics and electromagnetism. We are finding that some of the mysteries he and Faraday exposed at the molecular level are still mysteries today, but they are mysteries now more at the level of the atomic nucleus and surrounding cloud of atoms."

"I agree with what you say about Maxwell and Faraday," Thomas said, "but what can you tell me about the math and computation problems you are having at the subatomic level?"

Feynman stared at Thomas for several moments. Thomas was calm under this close scrutiny. "Do you know any magic tricks?" Feynman asked.

Now it was Thomas' turn to stare. "Is this a question about whether or not I believe in black magic; whether or not I believe in the scientific method; or whether or not I might prefer to believe in voodoo over mathematics?"

He frowned as though he had been profoundly offended. Then he looked up at Feynman to see if he could read his intent.

"It's a simple question," Feynman said. "Give me your honest answer so we can move on."

Thomas thought for a moment, then raised both his hands in front of his face and waggled his fingers at Feynman. "Booga-booga."

Feynman's bursts of raucous laughter could be heard across the room.

From *CHAOS*...

Page 295. Top of the page.

Once in the dining hall I took my C-rats to a table where five grunts were talking and playing a noisy card game with lots of cursing and slapping the table. Their rifles were stacked against an adjoining table, but the thing that convinced me that they were combat infantry were the Airborne shoulder patches on their jackets.

"Hey guys. Mind if I join you?"

"Shit yeah, man," one said. "Sit down. You got money or drugs?"

"Ah, no, but I do have a story to tell about how I got these wounds in a firefight on Route 9. Is that good enough?"

The hand slamming and bullshit ceased abruptly. All eyes turned to-ward me.

"Who you shittin', man? You're not even military. You're a correspondent hack trying to pick up some local color for some small town rag where everybody hates the baby killers, right?"

I shrugged. "Yeah. Basically, you're right about why I'm in this shithole country, but I'm not here in front of you hoping for some bullshit story for the home folks. I am looking for updated information on the conditions between here and Lang Vei. I got wounded in a firefight on Route 9 a week ago. I left a buddy back there, and I mean to go back and get him. Can you guys give me some help or not?"

"C'mon, man. Take the chip off your shoulder," the main instigator said. "Call me Hack. If you are worried about military rules you can call me Lieutenant Hack. You want a toke?"

"Sure. I'll take a drag." Hack passed what was left of a joint over to me. I took a long pull on the roach. I did manage to burn my fingers on it, but the charge felt good going down.

PREAMBLE

The concept that is driving the creation of this book began several years after seeing the 1971 Australian movie **Walkabout.** The movie is taken from an award-winning book of the same name published in 1959 by James Vance Marshall. The book was re-released by the New York Review of Books in 2012.

The movie's director, Nicholas Roeg, lays the opening scenes in a modern Australia city, probably Sidney. We are introduced to an upscale suburban white family. There is a man in a tie, probably the father, who seems troubled. There is a post-puberty private school girl and her much younger, sun-bleached blond brother. Both children are dressed in the pressed uniform and hat for their school.

While these early scenes unfold a young Aboriginal boy begins his tribal rite of passage deep in the outback. This is his walkabout - the concept behind the movie's title. He will have a bow, some arrows, a knife, a "possibles pouch" of tools, charms, and totems for the unexpected events that may challenge him during his journey of weeks, perhaps months, across the parched and trackless desert.

The essential story begins as the father drives deep into the outback with his children for what might be a picnic. Suddenly, an event too terrible to discuss here leaves his two children without their father. In an instant they are abandoned, traumatized, and frantic.

After a few days of aimless wandering, desperate to find water, their school uniforms dirty and ripped, their paths cross with that of the black, nearly naked Aboriginal. The Aboriginal knows how to find water, but he speaks no English, and the cut-up bloody parts of a large lizard hanging from the decorated piece of woven cloth that passes as a waist band is gross and disgusting to the high-born children.

He has no experience of the pale and frightened children before him, nor of their parentage. Their strange dress tells him nothing of their tribe or clan. He cannot understand why they are reluctant even to touch the few mouthfuls of muddy but drinkable water the Aboriginal has found in a damp spot at the bottom of a shallow depression.

This collision and merging of two vastly different cultures, in all their gritty, grisly, and conflicting detail, has become a powerful metaphor, for me, for all cultures in all times and in all places.

The seeds of this concept began to germinate during my service in the U.S. Peace Corps in the late1960s. I was the only non-Afghan member of an Afghan land survey crew deep in the deserts of Afghanistan. Roeg's movie gave my concept of clashing cultures the kind of clarity and focus needed to sustain a book writing project of this size and scope.

In **Cassiopeia's Quest – Revelation** the churn of merging cultures and the story of Queen Cassiopeia herself are the binding force that holds these 10 independent stories together over the 100,000 years of their chronology.

Narrative technique is also important. I am a constant reader and regular movie-goer. The structure, visuals, and story-telling in books and movies has guided my thinking - and my aesthetic - throughout the years-long creation and writing of this book.

The other part of the skill set required is the actual writing. Hopefully, my Bachelor of Arts Degree in English many, many years ago – along with the help and support of my many readers, like my art accompanist Hetty Barthel and my professional editor, Jessica Hatch, and the style of some of Hemingway's finest writing - will now help me carry that load.

The "Personal Note" and the "Reader's Guide" sections in the back pages, have more on these topics.

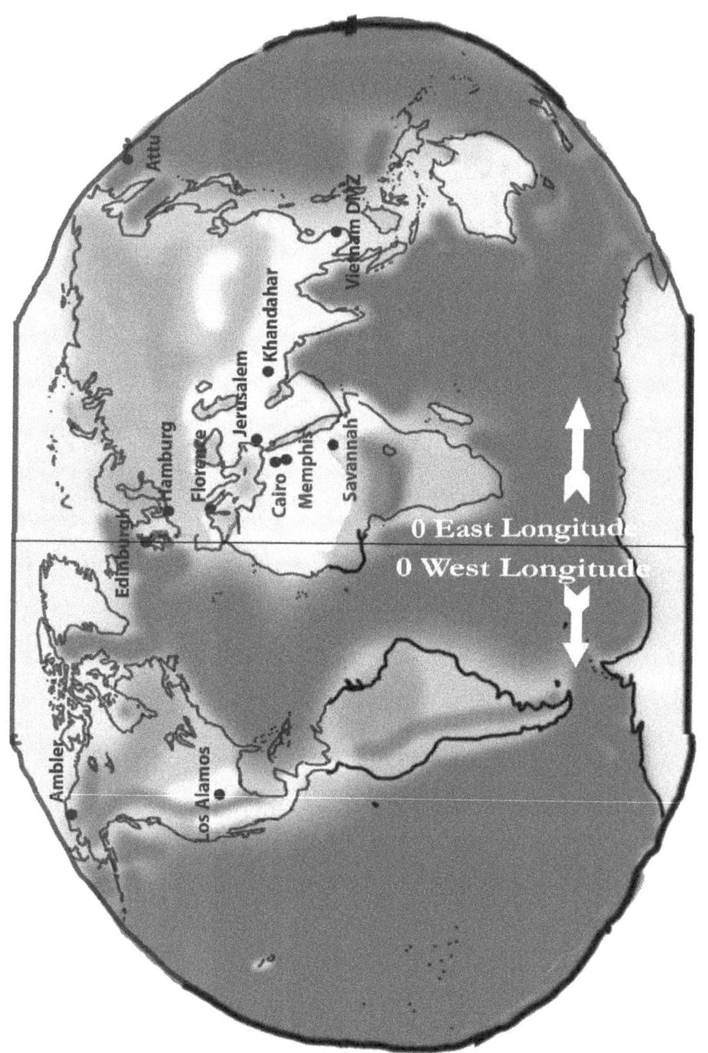

Map Of The World

This is a Map of events in the world of **Cassiopeia's Quest – Revelation.** It is in a way, her Walkabout.

Cassiopeia: Queen of Aethiopia

Ten original short works of historic fiction are offered in this book. Each story is particular to a time and place in the culture of homo sapiens as humanity evolved in their descent from the enfolding embrace of the trees onto the wide horizons of the grassy savanna and into modern times. The events that occur in that time and place in each story are generally well known in history. I am not a historian, but I have tried, with my lifetime of voracious reading and research, my pushy curiosity and my extensive web-based searching, to portray historic events with as much accuracy as I can find. The 10 stories are presented in chronological order covering a little over 100,000 years.

Each story is centered on a young, intelligent women as protagonist. She is a fictional character; as are most of the characters directly associated with her in the story. She is a woman who has passed puberty and is now starting to assert herself in her world. In the time and place of her story, she is beginning to know who she really is and what she really wants to do.

This collection and these stories are, in many ways, my homage to the strong women I know and have known in my life including the women in my extended family and among all my closest friends. These are women with edges and opinions, and they are usually not shy about expressing themselves. They are smart and they are funny. They are socially aware and socially engaging. I'd like to think that I possess the same strong character as these strong women.

The name of the female character in the title of this collection, Cassiopeia, Queen to the King of Aethiopia, Cepheus, may be the strongest woman in all human culture. She dominates the night sky in the northern hemisphere. Her constellation lies only a few degrees away from the most important star in the modern night sky because of its importance in celestial navigation, the earth's pole Star, Polaris. Cassiopeia resides there in all her majesty and queenly glory. Her immediate family: King Cepheus, her daughter the Princess Andromeda, and Andromeda's husband the mortal warrior among the gods of Greek myth, Perseus alongside his horse Pegasus. All are with her.

Queen Cassiopeia has come down steadily and in something close to her present narrative form from her origins in the deep mists of time.

From a period possibly long before the time described today as prehistory, she may have been a real queen. By one account her story as embodied in her namesake constellation - as recorded in human history as a recognizable object in the night sky - has origins in Babylonian Astronomy perhaps five thousand years before the present day.

The events from which the story of the title character may have been taken could even have occurred much earlier. Perhaps around the time that upright, two-legged, ground-bound homo sapiens humans had begun to learn ways of expressing themselves beyond the grunts, shrieks and animal mimicry more typical of their tree-bound ancestors. Perhaps when they began to look up toward the night sky they began to wonder, with the members of their clan or tribe in their primitive way, if there is some meaning in the patterns of the pinpoints of light above them that might explain their mysterious world to them.

The first story in this collection, Savanna, describes an incident in the lives of a small clan of hunter-gatherers. In their evenings they may have struggled to describe the images they might have seen with the more intense long-distance vision needed in the grasslands or as they might have imagined when they tried to fill in the gaps between the stars and other bright objects in the night sky 100,000 years ago.

By the time of our first story those hunter-gatherers may have been desperate in their quests for a greater understanding of all aspects of their world. Their world had begun to take shape for them by the expanded senses and memory that came along with their evolutionary descent from a life and a world lived mostly enfolded in the tops of trees to a life and a world lived mostly on the open mostly treeless grasslands of the Savanna.

The story of Cassiopeia and her royal Aethiopian family is a simple one though horrific in some of its mythic details. It is mostly consistent from the beginnings of Greek mythology as it evolved, in turn, from roots in oral storytelling and singing carried by minstrels and vagabonds from town to town in ancient times. These oral stories evolved into the writings of one or more writers now known to us as the one Greek writer, Homer, in the stories of the Illiad and the Odyssey; stories that became known almost 500 years before the time of Alexander the Great of Macedon, a character in the third story, here, titled: **Empire**, and 800 years before the time of Jesus of Nazareth, a character in the fourth story, here, titled: **Faith**.

The story of Cassiopeia is generally told this way: Very beautiful and vain, the mortal Queen Cassiopeia committed hubris by saying that she and her daughter the Princess Andromeda were more beautiful than the 50 daughters of the sea god Nereus, who called themselves Nereids. In reaction to this offense, Poseidon – god of the Oceans and brother to the King of all the gods, Zeus - became infuriated. So infuriated that he sent the mythical sea monster Cetus to plague the coasts of Aethiopia. The King and Queen then consulted the Oracle of Ammon (an Oracle we will hear about again in the story of the Macedonian leader Alexander the Great). They were told by the Oracle that they had to sacrifice the beautiful Princess Andromeda in order to appease the wrath of Poseidon, god of all the oceans.

By one account the King chained their daughter to a rock next to the sea as a sacrifice to Cetus, the sea monster. Perseus – returning from Titan astride his great horse Pegasus after slaying Medusa with her severed head of snakes in his hand – saw Andromeda chained to the rocks and saved her by forcing Cetus to gaze upon the severed head of Medusa, thus turning the sea monster to stone. Following this heroic act Perseus then fell in love with Andromeda. With the blessings of the Aethiopian King and

Queen Perseus then proposed to Andromeda. They married and bore nine children.

Furious that the mortal Queen Cassiopeia was not yet punished for her sin of arrogance and offense to the gods, the god of the Ocean, Poseidon, tied her to her throne and put her in the heavens near the earth's pole star close to her daughter, to Perseus and to the King. She would revolve upside down half of the time and in that way be humiliated for all time, at least in the mind of Poseidon.

To me, however, Cassiopeia has never been a pathetic figure to be humiliated by a mere god. Though she is upside down on her throne for about half the year in the mind of the sea god Poseidon, she is upright on her throne with all her family close by for all the year forever.

The consolidation of stars and myths into 48 Constellations, including Cassiopeia, was first prepared and documented in a commonly understood form by Claudius Ptolemy, the last living member of the Ptolemaic dynasty of Egypt. Ptolemy published his catalogue of constellations – his **Almagest** - during his many years in study at the Great Library of Alexandria.

This around the year 150 in the Common Era (CE) after the time of Christ. One source says that Ptolemy's work was drawn from a star chart drawn 500 years earlier by the Greek Eudoxus of Knidos in 350 BCE.

Other sources claim that the Babylonians, as early as 3000 BCE, prepared a chart of Constellations and recorded them on a stone tablet using a system of wedge shaped forms pressed into wet clay with a stylus. The Babylonians prepared their charts in a language developed in Mesopotamia called cuneiform. Once dried these clay tablets have become a permanent record in stone able to be passed down the generations to the present day.

I am prepared to assume that Cassiopeia was a real queen deep in the mists of time, and, because of her great power and reputation among oral story-tellers and minstrels over time, one of those represented among the Constellations in the Babylonian Chart of Constellations.

***

The full title of this collection is Cassiopeia's Quest – Revelation. In the modern catalog of 88 constellations and the associated stars within their boundaries - as adopted by the International Astronomical Union (IAU) in 1928 - there is no mention of Poseidon.

All that is left of my story of Cassiopeia are the 10 stories in this collection.

I hope that all my readers will experience revelation while my young protagonists live, work, play and wonder at the meaning of the world under the influence of Cassiopeia, Queen of earth's northern sky at night.

I hope you and all my readers enjoy this book.

Jerry Smetzer, Writer and Painter

November 2019 in Juneau, Alaska USA

Savanna

Spring, 100,000 BCE. The east African savanna. Ah'-oom, a talented hunter, will lead an expedition along a river toward a herd of beasts; at least one of the beasts must be taken for the meat, fat, bone, hide, and sinew needed by the members of her small clan of hunters and gatherers.

====

Ah'-oom has been headstrong and energetic since her birth many seasons ago. As she seeks to master all of the elements of her environment and her complex relations with other members of her tribe, she has learned many lessons. With the help of her mother, Yah', she learned how to make her own knives out of bone for cutting meat, and her own needles to sew animal hides for warmth. She has made a stone hammer to subdue game needed for food.

As she grows more skilled and aggressive in the use of her tools, and as she begins making spears at the request of the hunters, she grows bold enough to tell Boa, the lead hunter and the father of Ah'-oom, that she wants to go along on a hunt.

Boa' has tried many times to stop her from going on the hunt, but Ah'-oom, as headstrong as he, goes along anyway, though usually at a distance from the main hunting party. Eventually, though, Boa notices her strengths on the hunt – her talents for tracking and striking game; for paying close attention to her surroundings and for her way of calling out the dispositions of her hunting partners.

As Ah'-oom gains trust among the other hunters, she learns that she must adhere closely to the overall goals and purposes of the clan hunters as a group in order to increase their success in bringing meat, bone, hide, and sinew back to the other members of the clan.

Ah'-oom is physically small, but her vision of what must be done on the hunt is strong and clear-eyed. When she comes into conflict with those who resist her decisions, she must prod them, sometimes with one of her weapons: the spear, the bone knife, or the stone hammer. One that she cannot subdue with prodding is Ar-gah', an abandoned child close to Ah'-oom in age who was taken into the group several seasons earlier has large patches of dark, bare skin on her back, chest, and upper arms and legs. She is more prone to discomfort during periods of cold than the other members of her group. She often needs to wrap herself in extra animal hides for warmth. Yah'-nay is careful to make sure that Ah'-oom has enough animal hides around her when the group huddles together at night.

***

Ah'-oom and Yah'-nay take pleasure each in the company of the other. Yah'-nay exhales sharply through pursed lips and points toward a smudge of stars that show clouds gathering in the northern sky. She lifts her nose and sniffs. Ah'-oom also sniffs, then motions for silence. She becomes alert. She stands and looks toward the source of the gathering breeze.

Ah'-oom points in the direction of the breeze and sniffs again. With hand movements and a guttural sound deep in her chest she indicates that several large animals are perhaps a day's distance away. She turns to grab her spear shaft and the piece of bone she has sharpened, the piece that will be tied to the end of the spear.

She notches the shaft to accept the bone, and she wraps shaft and bone with a strip of hide leaving most of the spear point and cutting edge clear. She returns to the group and kicks some of the others awake. She pulls three larger members away from the other tangled bodies. Those who object or resist, receive another kick or a sharp punch to the head from Ah'-oom.

Boa, her father and the senior male in the group, rouses himself and gives a low bark toward Ah'-oom. She looks up and stares at him. She is not sure if she has drawn his ire. In hopes of explaining herself, she points with jerking motions back to the breeze, so redolent of warm animal flesh. Her fear is unfounded.

Boa stands, stretches, and passes his stone ax, the symbol of his leadership, to her. Ah'-oom grips the ax handle with both arms. He motions her and the others toward the source of the freshening breeze, and to the several ravens flying in the same direction. Ah'-oom calls again to three of the group's members, its hunters, and points to the birds. It is an omen.

One of the hunters, Ar-gah', is an adopted member of the group who is thought to be two seasons older than Ah'-oom. He was taken in many cycles of the sun ago as an injured youth who had either been abandoned by his familial hunting group, or they were somehow injured or killed. Boa

was impressed that he had found water, had hidden himself from predators, and had survived for several days even though injured and unable to move quickly. Had Boa not seen value in the foundling, he may have left Ar-gah' on his own. No other member of the hunting group would have argued ===

Ah'-oom begins to understand that Ar-gah' is angry when Boa looks to her for some of the fundamental decisions to be made on the hunt. Worse, in the mind of Ar-gah', Boa does not criticize her severely enough when her lack of experience or excess energy puts one or more of the other hunters in danger.

Part 1.

The scattered, bright stars in the dark sky above the African savannah fade as the horizon in the east lightens. Behind a mound of earth, out of the wind, a small group of hunters huddles. bodies touching and arms and legs tangled together for warmth. They stir out of their sleep. Yah'- nay, the eldest female of the group, is already up and looking toward the brightening edges of darkness. Her view of the moon and bright stars in their various patterns above fades in the gathering light of morning.

Yah'-nay looks down toward the horizon. She sees that the sun, at the point of rising, moves farther north each day. She remembers that the shadows cast by the sun at midday will continue to shorten until, when the sun is highest in the sky, they will then begin to lengthen.

Yah'-nay knows from this that the spring rains that follow the cold rains of winter are also passing and that the days and nights, even the rain, will be warmer as the sun grows hotter in the days ahead.

Yah'-nay pauses in her remembrance of the sun moving farther south again as the air grows cooler until the shadow grows to its longest extent. She knows there will be changes in the moon from a thin sliver to a full round bright ball and back to darkness, then the sliver again. There will be several of these changes in the moon before the air begins to grow cool again in its continuing cycle of change from sliver to ball to darkness and then back to sliver again; ...a cycle that she has always known as the time between her bleedings when she was younger. A cycle now known to the beautiful first born child of Yah, Ah'-oom. Yah is mated with the first male child born of Yah'-nay's body, Boa. Boa is the oldest male in the group.

As she thinks about the things she sees in the sky and feels in her body, Yah'-nay struggles to make sense of the relationships between the regular movements of the sun, the moon, and the bleeding she felt and saw and touched and tasted when her body was younger.

She is joined by Ah'-oom, now two cycles of the sun past the start of her bleeding. Ah'-oom squats near the older woman and pulls the ani- mal hide away from Yah'-nay's shoulder. A purr deep in her chest begins as she picks the tiny vermin buried deep in Yah'-nay's thick body hair.

Several of the hunters have hair over most of their bodies, but Ah'-oom has large patches of dark, bare skin on her back, chest, and upper arms and legs. She is more prone to discomfort during periods of cold than the other members of her group. She often needs to wrap herself in extra animal hides for warmth. Yah'-nay is careful to make sure that Ah'-oom has enough animal hides around her when the group huddles together at night.

Ah'-oom and Yah'-nay take pleasure each in the company of the other. Yah'-nay exhales sharply through pursed lips and points toward a smudge of stars that show clouds gathering in the northern sky. She lifts her nose and sniffs. Ah'-oom also sniffs, then motions for silence. She becomes alert. She stands and looks toward the source of the gathering breeze.

Ah'-oom points in the direction of the breeze and sniffs again. With hand movements and a guttural sound deep in her chest she indicates that several large animals are perhaps a day's distance away. She turns to grab her spear shaft and the piece of bone she has sharpened, the piece that will be tied to the end of the spear.

She notches the shaft to accept the bone, and she wraps shaft and bone with a strip of hide leaving most of the spear point and cutting edge clear. She returns to the group and kicks some of the others awake. She pulls three larger members away from the other tangled bodies. Those who object or resist, receive another kick or a sharp punch to the head from Ah'-oom.

Boa, her father and the senior male in the group, rouses himself and gives a low bark toward Ah'-oom. She looks up and stares at him. She is not sure if she has drawn his ire. In hopes of explaining herself, she points with jerking motions back to the breeze, so redolent of warm animal flesh. Her fear is unfounded. Boa stands, stretches, and passes his stone ax, the symbol of his leadership, to her. Ah'-oom grips the ax handle with both arms. He motions her and the others toward the source of the freshening breeze, and to the several ravens flying in the same direction. Ah'- oom calls again to three of the group's members, its hunters, and points to the birds. It is an omen.

One of the hunters, Ar-gah', is an adopted member of the group who is thought to be two seasons older than Ah'-oom. He was taken in many cycles of the sun ago as an injured youth who had either been abandoned by his familial hunting group, or they were somehow injured or killed. Boa was impressed that he had found water, had hidden himself from predators, and had survived for several days even though injured and unable to move quickly.

Had Boa not seen value in the foundling, he may have left Ar-gah' on his own. No other member of the hunting group would have argued against it. As it turns out, Ar-gah' has grown into a good hunter, and his loyalty to Boa and to the welfare of the group is total.

Ar-gah' senses the affection that Boa feels toward him, and as such, this morning, he stares with a deep frown at the ax in Ah'-oom's hands. Has Boa not yet felt confidence in his hunting skills to give him leadership? What more must he do to prove himself worthy?

The other two hunters are Ah'-oom's uncles. Bo-nee' is a skilled tracker, able to find the tiniest and lightest footprints in the loose sand and grasses and in the twisted branches and crushed leaves in shrubbier areas. From the tracks, he can then detect the directions of travel of possible game.

Bo-nah' is the strongest of the group, able to throw a spear a long distance with deadly force and accuracy. He is also very imaginative with the materials used for tools and weapons, which Ah'-oom and Ar-gah' both admire. As he puts a tool or weapon together, Bo-nah' often pauses to look at the materials and think about how they might be reshaped and extended in order to improve his ability to throw with power and accuracy.

Bo-nah' knows that his value to the tribe lies in how well he can inflict a crippling injury from a long distance on the animals the group needs for food, warmth, and shelter.

As much as possible, Bo-nah' passes his skills along to Ah'-oom both because Boa has expressed his pleasure in it and because Bo-nah' takes pleasure in passing his skills along to such a bright, attentive, and capable pupil.

Today, at Ah'-oom's call and Boa's urging, Bo-nah' picks up his spear, and, with it, a long piece of thin leg bone he has been working at one

end with abrasive stone. He has abraded a pocket in the end of the bone in a way that can hold the back end of a spear shaft. He has not yet tried out his idea in a hunt, but the spear end fits well together in the pocket Bo-nah' has created. It appears it will, with practice, make the force of his throw more powerful and the accuracy more true. He hopes he will have an opportunity to use his spear together with the legbone pocket in the course of the hunt.

As the tribe readies for the hunters' departure, Ah'-oom's young sister Pe'-dee dashes up. In her small hands, she cups a river rock. Its vivid blue catches the light of the sun and reflects the vibrant sky above.

The group cannot resist. They pause what they are doing to marvel at the stone. When Ar-gah' reaches out for it, Boa bats him away. With care, he gently takes the stone from Pe'-dee and stands fully upright with it protected in his fist. The others understand his meaning. This stone is important and special, and as such, it will only be given to a member who has earned its beautiful blueness. Before they can leave, Yah'-nay scoops Ah'-oom close and holds her

against her chest. The women know the threat of other tribes that encroach on the savanna. Ar-gah' has seen them with his own eyes, from a nearby ridge. He imitated the way they loped across the ground the other night, in front of a fire. The way he jutted his jaw forward and menaced his brow had frightened the little ones. The hunters will not only have to look out for wild animals but will have to protect one another against the fear of these shadowy and possibly violent strangers.

Ah'-oom pulls back and jerks her head confidently at Yah'-nay. She is a skilled hunter. Boa is trusting her to lead his group. She will not let danger befall them.

Ah'-oom and the hunters move down to the creek that runs near their sleeping place. She urges them to drink as much water as they can stand and carry as much as they can in their skin bags. They stop to pick up their clubs, spears, and strips of sharpened bone, then start moving in the direction Boa has indicated. Ah'-oom picks up her father's ax and follows them.

The sky is light now, and only the brightest stars remain visible. The sun has not fully risen, but Ah'-oom, as leader, knows that the day will be hot and the distance to the game long. Travel will be over sandy, windblown desert and grasslands where water may be hard to find. She will have to stay alert for the smell of damp ground if the day turns too hot. Each of the hunters knows that they must call a break from the hunt whenever they find small animals, reptiles, insect mounds, or clumps of edible plants: they must be gathered and shared for food.

Once Bo-nee', the last still slaking his thirst, has had his fill, the hunters begin to jog at a pace that all of its members can maintain through the day. When the sun rises well above the horizon, Ah'-oom runs ahead to the top of a small hill and signals a stop. She shakes her arm in a way that signals her wish for absolute quiet. She sniffs, she twists and turns her body as she tilts her head from side to side, and she nods up and down with her hands cupped to her ears, listening for the normal sounds of birds, animals and rustling foliage. She sniffs for the smells of the game they are hunting. She looks in the direction of their prey for any disturbance among birds, or any animal activity that stirs unusual amounts of dust.

Then she stops, still, and squints to the north. Her shadow falls before her as it is cast by the southern sun. Before leaving the small hill, Ah'-oom looks back in the direction they have come. She has been memorizing significant landmarks by observing them as she approaches, then looking back at them after she has passed. She looks at the sun's course across the sky, and at the length and direction of the shadows cast by the hunters' bodies.

She feels, sniffs, and tastes the wind with her tongue, and looks at the movement of clouds. She looks back into the distance for the shapes around the group's sleeping place, and particularly to the large acacia tree standing near the windbreak that marks it. She will need to find and follow these shapes on their return if she is to find the place where her family sleeps.

Finally, she descends from the mound and jogs to the north. The other hunters follow, matching her pace. As they run within a few yards of each other, Bo-nah's mouth reshapes itself into a wide grin. He vents a high pitched, warbling cry of enthusiasm for the chase. The others also cry out as they run. Ah'-oom looks into the distance for the next landmark on their way north.

***

Ah'-oom has called the hunters to a stop. She has, once again, ascended a small rise to taste the wind. The sun is now past the peak of its travel through the sky, and they have not yet spotted the game that Ah'-oom has said is ahead of them. Ar-gah' is particularly disgruntled and makes it known in the way he noisily sucks on his water skin. All check their weapons and make small adjustments as though getting ready for the hunt itself, rather than just the possibility of a hunt sometime in the future.

Ar-gah' stretches, yawns, and scratches his torso as though he is losing interest in the main purpose of the hunt, and may have a different idea.

Bo-nah' recognizes his fading commitment and wonders what it might mean. He knows that Ar-gah' is a summer season or two older than Ah'-oom and somewhat more muscular, though he does not run as fast, cannot throw as accurately, and his trail sense is not nearly as well developed as Ah'-oom's.

More important than that, Boa has given the leadership of the hunt to Ah' -oom. The implied message to Bo-nah' is that he needs to remind Ar-gah' of this simple fact. He reminds the boy of this with a series of high-pitched barks, staccato grunts and yips close to his face followed by a sharp poke to Ar-gah's chest. Ar-gah' backs away and turns to continue checking his weapons in preparation for the continued chase to the north.

Bo-nee' is loyal to Bo-nah' and will support whatever his brother chooses to do. That, he knows, as he stands staring at the two posturing male hunters, is to honor the wishes of Boa. Even so, Bo-nee' likes the impulsive Ar-gah', and the two are often seen together as he teaches Ar-gah' the skills of tracking.

Ah'-oom lets out a low whistle followed by a series of chirps. The others turn toward her. She is mimicking the cry of a bird common in this area, but the three hunters know that this call means that she wants Bo-nee', their expert tracker, to join her on the rise. He jumps up with his spear and trots toward her. Once he joins her, Ah'-oom points to the ground in front of the rise, then to the copse of trees that lies a ways beyond. Bo-nee' sees slight disturbances in the soil in front of them, and a line of shadows in the trees as though something has passed this way and gone into the woods. He nods to Ah'-oom. The game is close by.

She turns to Bo-nah' and Ar-gah', touches her tongue to the roof of her mouth, and gives a series of five clicks, a pause, then five clicks again. It is a message that they need to approach her, but silently. As they gather around her, she motions Bo-nee' to move off the mound, tracking the signs in the dirt and vegetation toward the shadow in the trees. She then motions for Bo-nah' and Ar-gah'' to follow him as quietly as they can. All are crouched low. Ah'-oom waits, then moves to the right and upwind of the other hunters. She checks the ax kept in the hide belt around her waist, then makes sure, once again, that she is correctly holding her spear.

None of their spears are perfectly straight. Even so, with practice and Bo-nah's help each has learned how to hold and throw in such a way that they hit their target with as much force as the hunter can bring to the throw. The purpose of the spear in the hunt is to make a killing strike that will cause massive bleeding. Failing that, the hunter needs to be close enough to the animal or bird, to cripple in some way its ability to outrun the hunters and escape. If the animal is close enough to strike without the need to throw a spear, then the hunter may be able to shove his spear into one of the animal's killing zones with great accuracy.

Failing an immediate kill, the hunter may be able to pull his spear out of the animal and try again. Such opportunities are rare, however, as are the hunters with enough calm, skill, and experience to know exactly how to place a killing blow in a particular animal, especially when that animal is charging the hunter, or threatening him or her with a snarling growl and bared teeth.

Such an animal – a very large cat-like animal common on the savanna - though not yet snarling or charging, appears and faces Bo-nee' from the edge of the woods. Bo-nee' stiffens but dares not make his excitement known. The wind is blowing toward the hunters from the woods, so the animal cannot yet smell or hear them, but it can see unusual shapes in the distance. Those shapes, the hunters, have stopped moving on Bo-nee's signal. They are now waiting silently and without movement to see if the animal will move in a direction that gives the hunters a real opportunity to take him. Ah'-oom, several strides away from the others, has also spotted the animal. She silently takes in the situation.

The edge of the woods where the animal now stands is about as far away as any one of the hunters, even Bo-nah', could possibly throw a spear. At such a distance, accuracy would not be much of a consideration. The most likely result of such an attempt would be a scared animal that runs back into the woods. As that thought crosses her mind, Ah'-oom jumps, screeches, and runs toward the other hunters. The animal moves quickly. Just as Ah'-oom has planned, the animal perks up at this disturbance, and leaps for the running girl.

Bo-nah', Bo-nee', and Ar-gah' remain crouched, but they grip their spears more tightly as they estimate the shrinking distance between them and the now charging animal. They know from Bo-nah's training that the distance will have to shrink to less than twenty strides before they can risk throwing a spear. Otherwise, they might lose the spear without diminishing the speed of the animal's charge, and their tools for defense will be gone.

Ah'-oom runs quickly toward the other hunters while carefully watching the closing distance between the animal and the others. Without warning, she jerks to a stop, turns, and runs as fast as she can away from the hunters. The animal is hot on her trail and turns with her, thus presenting the crouching hunters with the full length of its body.

Ar-gah', ever the impulsive one, is first to leap up and throw. His spear hits the animal near its hindquarters but glances off, leaving only a minor cut that will neither slow the animal down nor bleed enough for the hunters to be able to track it. Ah'-oom continues to run; her life, and the lives of her hungry tribe, depends on it. Bo-nah' snarls and chatters at Ar-gah' before throwing next. His spear flies true to the animal's chest but does not pierce the heart. Even so, the spear pierces deeply enough to cause the animal to stumble in its run and fall tumbling and sliding to the ground.

The three hunters run screaming, victorious, with their weapons toward the struggling beast. Ah'-oom, on hearing their yells, looks back, then turns and runs toward the animal with her spear and her ax at the ready. The beast struggles to get up. Still several strides away, Bo-nee' throws his spear in the hopes that he can strike a death blow. The animal only roars in pain.

Bo-nah' runs at top speed with his stone hammer. He literally flies to the beast and brings the hammer down with all the force he can muster on its head. But, in that instant, the animal turns toward the approaching hunters and tries to bat down the object flying toward him. His claws catch Bo-nah's throwing arm and deflect the force of his hammer. Bo-nah' cries out. There is no time for Ah'-oom to check on her uncle. The three other hunters are on the beast with their sharpened bone knives, tearing into its vital organs through the soft tissue of its belly.

The beast squeals. It tries to bat them away, but its blows are now feeble.

Eventually, its life force drains away, and it lays back to die.

All of the hunters scream in pain and agony. All break down in long, sobbing cries at the release of the tension from the killing. In the animism that articulates the hunter's holistic view of the world around them they have killed a member of their immediate family. Their grieving is all the more intense for it. All the others have cuts where the animal's blows caught them in his death struggle, but Bo-nah's arm is injured seriously.

Ah'-oom, teary-eyed, bends to the dead animal and places her hand over the animal's still open eyes. After a moment of grieving, Ah'-oom, herself bleeding from a cut in her scalp and on her hip, attends to the others before applying a poultice, prepared and given to her by Yah-nay before they left the sleeping place, to her own wounds.

Apparently, Bo-nah' has suffered no broken bones, but he is woozy and unfocused from loss of blood.

Ar-gah' is the least injured, a fact that, considering how impulsively he threw then lost his spear putting the other hunters in great danger, aggrieves Ah'-oom. She directs him to retrieve his spear, and then check all their weapons for any damage that needs to be repaired. The beast may have had a mate or been associated with a large pride. Worst of all, the smell of blood might attract any number of curious beasts, like jackals, to them.

Ah'-oom worries her lips with her teeth. Bo-nah' will need some time to recover from his wounds if he is to be any help in defending them from animal attack and getting them all back to their home territory. She knows they will have to spend the night away from the sleeping place. Her mind fills with all the problems that will need to be solved if they are all to get back home safely. Most importantly, she wonders how, without their strongest hunter, they will carry the hide, the bones, and the meat of the beast that has caused all this grief, back to the others.

Ah'-oom turns again to the beast and puts her hand on its bloody chest.

In a low voice she hums a song of respect to the animal for giving itself to their hunting party. In her mind she promises that all of its parts will be used to help with the important work of the hunting family's survival.

She looks toward the woods and contemplates the possibilities for shelter among the trees. Since she and her family have little experience among the trees, she also wonders about the risks of animals sneaking up on them in the night. Ar-gah' finishes his work recovering and repairing the weapons. Ah'- oom directs him to go into the woods and tell her if there is a safe place to move into for the night. He needs to find some tall trees, some downed trees, and some binding materials like vines. These can be assembled into barriers to protect them and the salvaged parts of their killed animal, now buried to cover the smell of blood and rent flesh, from the beasts that will be roaming in the night.

She gestures toward the surrounding grasses in a way that tells Ar-gah' to gather bunches of it for bedding and warmth, then motions toward Bonah' and his need for restful sleep. They will need to start their journey back to the sleeping place very early in the morning if they are to make it back there by nightfall.

In the gathering dusk, Ah'-oom follows Ar-gah' into the woods to inspect his fortifications in the area where they will try to give Bo-nah' some rest. She is concerned that animals will attack them in the dark, unless they can get off the ground and lodge some sharp sticks pointed outward for defense.

Before she leaves, she goes to Bo-nah's side. He is awake but still groggy from his wound. He assures Ah'-oom that he can do his share of the work. She places a warm hand on his uninjured shoulder, and he turns back to the carving and shaving of the bone he will use to give himself a stronger and more accurate throw.

Part 2.

Ah'-oom walks with Ar-gah' toward the area that he has fortified for the night ahead. Ar-gah' points toward the wood he has piled around a tree big enough to support whatever they decide to build. Ah'-oom frowns at this pile, then looks at Ar-gah' and grunts her displeasure. She points to the sun sinking beneath the horizon as the evening skies darken into night. She motions for him to come with her to plant the larger branches around the base of the tree, leaving enough room between the tree trunk and the branches for each of them to sit with some protection from the animals of the night.

Ah'-oom then points to the smaller sticks scattered around and directs Ar-gah' to gather enough wood for a fire that will burn through the night. Once she and Ar-gah' have set some of the branches against the tree, she invites Bo-nah' into the shelter to rest. Bo-nah' seems to be recovering quickly, but he is glad that Ah'-oom did not give him any duties. He takes the piece of leg bone into the shelter and uses a small rough stone to continue shaping the spear pocket into a sort of hook at the end of the leg bone. The hook will hold the spear and leg bone together through the hurl and release. The hurl and release is the most difficult part of his idea.

Though each movement of his wounded arm—his throwing arm—causes intense pain and sweating, he is desperate to test his idea, even if he has to show one of the other hunters how to use it. Of the three hunters in the group, Bo-nah' rejects Ah'-oom. She will have other things to think about. He rejects Bo-nee' because his tracking skills will be needed to find the shortest way back, before his arm becomes a vulnerability, before the meat can spoil.

Ar-gah', he decides with a sigh, is probably his best choice. He knows that training with him at such a critical time will give Ar-gah' a feeling of superiority over Ah'-oom. With the skill and more effective hunting weapon that Bo-nah' could give to him, Ah'-oom, because of her small size, could never throw farther than he could, given the same tools.

It is too dark to see his work, but Bo-nah's sensitive fingers can feel the imperfections in the pocket he is carving in the bone. He uses the stone to gouge and smooth further. The others had foraged as far into the woods as possible until it had become too dark. Now, Ah'-oom motions to Bo nee' to build a fire and stack the wood nearby.

Ah'-oom knows that none of the others will fall into a deep sleep while on a dangerous hunt. She always finds it difficult to sleep away from her mother and father, away from the warm and protective huddle of the group. Bo-nah' is a possible exception, she thinks, because of his wounds.

Bo-nah' moves to the side away from the fire where the protective branches are thickest. Ar-gah' moves next to him. Ah'-oom and Bo-nee' sit on either side of the opening the better to watch the fire, and to watch for any movements in the woods beyond. Bo-nee' and Ah'-oom fall into a light sleep.

Once it is smoothed to his liking, Bo-nah' puts the leg bone into Argah'' s hand and then moves his hand over the spear pocket he had been working on. Ar-gah' grunts. He knows immediately what it is. He takes the bone from Bo-nah' and, unable to resist, he waves it in the close confines of the protected area. It is as if he cannot wait to try out this newest weapon.

Bo-nah' is pleased. He motions for Ar-gah' to sleep, then motions toward the sun's rising and touches the younger hunter's shoulder. Bo-nah', the effect of his wound catching up to him, falls into a deep sleep. Ar-gah', though, is alert with excitement and anticipation.

He picks up the legbone and a short branch that lies nearby. He runs his throwing hand over the bone and imagines the release of the spear. He continues imagining and practicing the throw and release of the spear without letting the bone drop for a long time.

***

Night passes without incident. The hunters are able to sleep through the night, mostly undisturbed. As the gray light of morning lightens on the eastern horizon, Ah'-oom rises from her sleeping place and stumbles toward the now nearly dead fire. There are enough coals to get the fire going again, and she does so, hoping to find some berries or leaves suitable for making a warm sweet drink to help them on their way.

Looking toward the path they would take back to the sleeping place, she notices a raven huddled against the morning cold on a tree branch a few paces away. The raven takes note of her, and utters some deep, rumbling vocalizations. Ah'-oom notices, with a smile, that the raven has one white feather on his right wing. She recognizes the bird as one of the many who visit her good friend Yah'-nay most days. It is as though Yah'-nay has sent her this bird to say that all will be well. Ah'-oom takes this to be a very good omen for their return trip.

She goes to wake the others. They begin assembling and organizing the weapons, the meat, and food supplies they will need for their return. Bo-nee' separates two of the longest branches out of the woodpile.

These will be used to carry supplies, and, if necessary, Bo-nah'. When Bo-nah' comes out from among the protective branches around the tree, however, it looks as though he had completely recovered. The others are grateful for this. Bo-nah' gestures that he is ready for the trail ahead, and that they should get started.

Part 3.

Before they can leave, Bo-nah' wishes to make sure Ar-gah' is trained on the new spear. Ar-gah' appears, smiling and carrying the leg bone that Bo-nah' has been working on for several weeks.

Bo-nah' picks up his spear and offers it to Ar-gah'. Ar-gah' turns away from the group and points at a tree about twenty paces away. He turns toward Bo-nah' and takes the offered spear. He holds the legbone in such a way as to fit the back end of the spear to the notched end of the bone behind his throwing hand. With his free hand he places the spear along the top of the bone, then grasps it with the same hand he uses to hold the legbone. With one hand he lifts the spear and bone together and cocks his throwing arm back as though preparing to throw.

With one quick move, Ar-gah' launches the spear. It misses the tree he aimed at by a little, but it completely pierces a smaller tree a little farther out and to the right of the target. Ar-gah' turns back to the hunters with a big smile across his face. Bo-nee' and Bo-nah' whoop and shriek with pleasure. Bo-nee' runs to Ar-gah' and tackles him to the ground in celebration.

As they roll around, Bo-nah' wants badly to join them, but holds himself back rather than risk further harm to his throwing arm.

Ah'-oom, though, jumps into the rolling pile of bodies and gives as good as she gets when the punching, kicking, and screaming begin in earnest.

Eventually, out of breath, she gets up, squats on her haunches, and rests with her head down while she catches her breath. Her wound from the beast the day before has begun to hurt. Rather than dwell on the pain she shouts and gestures that it's time to go. The four hunters gather their gear and load the pallet prepared by Bo-nee'. Bo-nee' had assumed that Ah'-oom would direct Ar-gah' and him to carry the pallet, but she moves forward to take one end.

She gestures instead that Ar-gah' should go ahead of them, and practice walking, running, and throwing the spear with the leg bone. She shows how he should always keep the group in sight. She points toward the sun and indicates to Ar-gah' that he should stop and wait for them when the sun has moved a certain distance. She can foresee many possibilities for violence on the trail ahead that might require the use of the new weapon, and she wants Ar-gah', now the strongest thrower among them, to be ready for all possibilities.

She sets a pace that will be best for Bo-nah'. Though he has protested that he feels fully recovered, she walks up to him, puts her hand on his arm, and looks deeply into his eyes, as though seeking the truth. He recognizes her determined look and looks away to the path ahead. Satisfied, Ah'-oom returns to the head of the pallet, picks it up in unison with Bonee' and motions for the others to lead the way. With Ar-gah' and Bo-nah' ahead of her, she will be able to assess Ar-gah''s progress with the new weapon and watch Bo-nah' for any signs of fatigue.

They make good progress as the sun breaks away from the horizon. The day before had given them surprisingly good weather, and this day looks good as well. Two days in a row at this rainy time of year are a needed and pleasant surprise.

As Ah'-oom looks up to check the clouds and weather, she sees a raven high above them. She cannot see if it is the raven with the white feather, but she continues to believe that its presence is a very good omen for the success of their return to the main camp. She points out the raven to Bonee', gesturing and speaking Yah'-nay's name to indicate that the bird was her friend and was watching out for them. By midday, when their body shadows are shortest, Ah'-oom knows she and the others will have to stop. Carrying the pallet is hard work, and the weight of it is cutting into her shoulders. Besides, it is time. The sun has reached the point in the sky that she had indicated to Ar-gah'; he is already walking back toward them with the leg bone and practice spear.

As Ar-gah' approaches, he seems very pleased with himself. Ah'-oom grunts, encouraging him to show them his progress in this newfound skill with Bo-nah's weapon. Confidently, Ar-gah' stands and points at a bush at least fifty paces away. He settles the spear into the pocket Bo-nah' had carved in the leg bone, then cocks his arm, ready to throw. With what seems like a flick of his wrist, the spear flies higher and faster than any of the other hunters, even

Bo-nah', has ever seen. It flies directly to the bush, pierces the middle of it and plants itself deep into the ground behind it.

A bit of dust stirs up, but the watching hunters think it might be a puff of wind. Ar-gah' walks to retrieve his spear and picks up a rabbit that the spear had penetrated. He does not admit whether he had known the rabbit was behind the bush before he selected his target. Though too tired to join each other in a screaming, shouting wrestling match, the three waiting hunters utter a series of sharp barks and howls to indicate their great pleasure at the progress Ar-gah' has made and particularly in the surprise and welcome gift of fresh meat.

The hunters strip the skin off the rabbit and eat the meat raw. There is no time for the social pleasures of preparing a fire and cooking. Then they covered themselves from the sun with hides so they can take short naps.

Bo-nah', once again, offers to take Ah'-oom's place carrying the pallet. She waves him away until she has napped. Before covering her head she notes that the raven has flown away to the west, and has not returned. This concerns her, but she refuses to share her worry with the others.

Some time passes, and Ah'-oom jerks awake. She goes to the other hunters to kick them awake. She barks at them to wake up and get moving. She does not want to lose more daylight and be forced to spend another night away from their sleeping place. All are groggy, but they assemble their gear and begin to move. Bo-nah' blocks Ah'-oom to prevent her from picking up her end of the pallet. She does not complain.

She notes that Bo-nah' seems to have fully recovered, though he does give out a yelp of pain when his wounded arm picks up his end of the pallet. Nevertheless, he motions her ahead to act as Ar-gah's eyes and ears along the trail. In addition to Bo-nah's leg bone, Ar-gah' has three of Bonah's best spears, leaving four for the other hunters. Ah'-oom still carries Boa's ax and her own spear and short bone cutting tool for defense. If they find some spear-size branches along the way, they will stop to make at least a few more spears. They soon come into some broken country where water sometimes flows after heavy rains. Judging by the dark clouds on the horizon where the sun normally rises, the possibility of flowing water, perhaps a flood of it, becomes very real. There will surely be animals, and there might be other hunting groups to deal with. They keep moving to the south with deliberate speed. Bo-nah' is as good as his word and does not slow the group down. He does not whine about the weight of the pallet, even though it is clearly causing him pain.

Ar-gah' moves quickly, and it is all Ah'-oom can do to keep up with him. After a time, she has to call a halt to wait for the other hunters. Ar-gah' points to a nearby ridge, suggesting that he will keep looking around the area. Ah'-oom continues to be grateful for his energy and enthusiasm in watching out for all of them.

The two hunters with the pallet join her and pause as she turns her body to look toward the horizon ahead of them; to sniff and taste the air; and to listen for noises that seem out of place. She hears the call of a distant raven, and though it calms some of her concern, she does not think it important enough to call attention to it.

Then, from nowhere, a new smell. This one is musky and strong and like nothing she has smelled before. Ah'-oom stiffens. She turns her head and body to the west where the sun sets. Her eyes, ears, and nose focus on something in the distance, and she gestures for absolute silence. She points. The two litter carriers strain to see what she does. Then, there, in the far distance, they see some shapes. Something about the shapes does not sit well with Ah'-oom. They move not like animals, but in the way that she and her fellow hunters do. She remembers the fears of other strange tribes as expressed by Yah'-nay and Ar-gah'. She needs to figure out if they are moving, and in what direction.

After some time, she looks at the other two hunters with a look of concern and questioning. Her uncles both point to the south—the same direction they are traveling. Ah'-oom swallows nervously. Their paths will eventually cross. She gestures that her uncles should stay very low but alert while she and Ar-gah' investigate these moving dots on the horizon.

She picks up one of their spears and balances it in her throwing hand. She indicates the other spears and that they should all be checked by the two hunters.

Ah'-oom moves toward Ar-gah'. He has seen the shapes and noted their movements. He is crouched and watchful. Ah'-oom crouches quietly beside him. Once again she moves her head and body to better get a feel for the atmosphere, the weather, and the movements of the figures. The figures are still moving toward the south, toward the peaceful windbreak and the unsuspecting remainder of Ah'-oom's tribe. Ah'-oom considers the possibilities of interacting with the moving shapes. They might be peaceful hunters, or they might be thieves looking for groups like hers with weapons and meat that can be taken away. Or they might be beasts like the one that had caused her hunting party so much pain and disruption.

If her group does not interact with them in some way, that group might be able to get to the sleeping place before Ah'-oom and the others. There are more of them than of Ah'-oom's group, and they appear to be moving more quickly across the ground. She thinks further. If the strangers are hostile Boa, Yah'-nay, and the others might need all the help they can get.

Ah'-oom hears a soft, warbling call from the raven overhead. She looks up and sees the raven appear to look back at her. The bird then turns west toward the strangers. She stands up and puts her hand on Ar-gah''s shoulder.

She looks in his eyes and indicates that they will approach the shapes in order to learn more.

Ar-gah' frowns at this. He stays crouched for a moment, as though to disobey her, but Ah'-oom slaps him hard on the head. She moves toward what now looks like a group of people who look like themselves. Ar-gah' growls under his breath and barks at her. He pauses to consider what Boa and the others in the sleeping place might do to him if he has to report that both Bo-nah' and Ah'-oom had been injured on the hunt, and he had not.

The killed animal he is bringing back might not look like much of a reward for their efforts on the hunt and particularly for the injuries they had suffered. There would be no songs of praise.

He then thinks that Ah'-oom could be seriously wounded or killed by the strangers. As it crosses Ar-gah''s mind, his body jerks, and he rubs his forehead vigorously to purge the thought. He gets up and runs to catch up with Ah'-oom. On his approach, she motions for him to stay a few paces behind. She points to the leg bone and spear, then indicates that he should hold the spear as a hunter would and keep the legbone out of sight. Ah'-oom knows that she and Ar-gah' will be outnumbered by the strangers. She hopes they will be seen as peaceful hunters rather than threats.

By the time they are within calling distance, the sun has arced closer to the horizon and night, and the dark clouds to the south grow toward them.

The strangers have stopped and now look toward the approaching hunters.

Ar-gah' holds his spear as a hunter would while looking for game.

Ah'-oom has taken the leg bone from him and put it in the strip of hide around her waist next to her father's ax. This way, she hopes it will look less like a weapon. Ar-gah' thinks grimly that he might have to quickly take the bone, fit it to his spear, and throw it with precision if their interaction with the strangers goes bad.

Ah'-oom walks to within a few paces of the strangers and stops. She gives two soft barks of greeting in their direction. She notes that there are less of them than the fingers on her two hands. All carry spears, but they do not seem to be threatening. Two of the strangers, males, carry a piece of hide wrapped around several long pieces of wood. The two males also carry a long, thin piece of wood with a thin strip of hide tied to both ends, forming a bow. She does not know what to make of this.

There are three women in the group. Two carry branches, which she imagines they use to build shelter when necessary. The third woman carries a hide slung over her shoulder that appears to be used to carry food stuffs. One thing about the group of strangers is very peculiar to Ah'-oom.

All of them except the oldest have very little hair on their bodies. She touches her own bare arm and feels comforted by this similarity.

Stranger still, when the other group speaks to each other, they move their mouths in ways that Ah'-oom feels she cannot. The others in her group are even less capable of making the sounds the strangers were making.

The strangers speak among themselves in very measured, rhythmic ways, almost like they are singing to each other as birds sing to each other.

Ah'-oom becomes fascinated with their speaking. She steps forward to more closely watch the way their mouths work as these sounds come out.

As she does, the beautiful sounds stop. She halts and looks up in time to see the two males stretching the bow and hides to aim the sharpened long sticks at her. Ar-gah' is by her in a flash, but she stops him before he can reach for the legbone. She motions for him to stay behind her. Ar-gah' continues to reach for the legbone, though, and Ah'-oom grabs her stone hammer from her belt and taps at his foot. He grunts in pain and bends to rub his foot where the hammer had hit.

With Ar-gah' calmed, Ah'-oom turns to the strangers. She puts both hands in the air, palms out, and bows her head to the ground in an act of submission. She kneels, still with her hands in the air, her head bowed.

She can hear the bowed hide stretch, and she hopes that this will show the strangers that they mean no harm. She grabs Ar-gah''s leg and rubs downward to indicate that he must follow her lead. Ar-gah' understands the need to show submission, but he only bows his head and opens his palms toward the strangers. He suppresses his anger enough to put one of his hands on Ah'-oom's shoulder to show solidarity. The oldest of the strangers is white-haired. He motions for his men to put down their bows and arrows. He approaches the hunters and invites Ah'-oom and Ar-gah' to rise. They do so and look into the eyes of the old man. He puts his hands on both their upper arms in turn, with Ah'-oom the first to receive his ministrations.

The old stranger motions to one of the women in his party carrying food. He picks some succulents out of her basket and offers them to Argah' and Ah'-oom. The two hunters accept them and eat them hungrily.

The old man then bids them all to sit and share their meager store of food. Ah'-oom tries, by pointing, to convey the existence of Bo-nah' and Bo-nee', and her need to rejoin them. Eventually, together, she and the old man agree that Ar-gah' will rejoin the other hunters, tell them what is going on, then help them get on the way back to the sleeping place.

Ah'- oom will stay with the travelers. As a sign of mutual respect two of the men with the travelers' party will go with Ar-gah'. They all know that they must then try to get to the sleeping place where there will be shelter and fire against the coming storm.

With gestures Ah'-oom urges the strangers to prepare for fast travel toward the hills to the south. She will lead them to the sleeping place. Once she has picked up her own tack, she picks up one of the bundles of hide that the others have brought to fend off the cold. Then, as at the beginning of her journey, she moves at the fastest pace that each of the strangers can match to return to her family at the sleeping place.

As the darkening clouds in the south grow more threatening, Ah'-oom hears the call of raven. The bird flies low over the group, appears to glance in her direction, then continues south.

They are on their way.

***

In the sleeping place, beyond the windbreak, Boa pulls the hide tighter around his body. A chilly wind has been building up ahead of the storm. He had been looking north and hoped he would not miss the returning hunters as soon as they came within his vision.

Yah'-nay joins Boa. She puts another hide around his shoulders and gestures to the new, partially buried shelter that she and Pe'-dah have built. Despite his pleasure Boa is growing more concerned as time passes and daylight fades. He fears that the hunters will spend yet another night away from their camp. He indicates to Yah'-nay that the fire needs to be started.

He hopes there will be enough wood and kindling to last through the night if that becomes necessary As they sit quietly together Yah'-nay's raven with a single white feather on its right wing comes to them and lands a few paces away. The raven hops sideways toward Yah'-nay then vocalize various sounds that she recognizes as sounds of comfort. She smiles to herself and chants softly in a way that gives assurance to Boa that all preparations are as good as they can be.

The fire is lit while there is still an evening glow from the setting sun.

Boa can no longer see very far in the direction of the hunters' return. He hopes they can see the fire.

The nature of his worry soon shifts. Shortly after, Ar-gah' appears followed by the other two hunters and two travelers bearing the palette with their weapons, and the animal they had killed the day before. Though he is glad to see three of his hunters return, Boa growls at the newcomers.

Ar-gah' does his best to cheer and bark in confident tones that all is well.

As soon as the hunters and travelers come in sight, the women begin to sing a barking ululation to indicate their great pleasure at their return. The younger ones run down the hill to greet them. Boa, Yah, and Yah'- nay descend more carefully, being mindful of the hazards on the darkened path. Boa is immediately concerned that Ah'-oom is not with them.

When Ar-gah' realizes that she is not among the greeters, he, too, becomes concerned. He and Bo-nah' turn back to the trail to search for the missing girl. Boa puts his hands on both hunters to caution them against going back into the dark, stormy night. He indicates that a better plan would be to pile more wood on the fire to give the missing party a bigger target in the dark.

Ar-gah' looks at the gathering storm and feels the rushing wind. He agrees with the wisdom of Boa. The possibility of the two groups missing each other, and becoming more lost, is very great. The group moves back up the hill to gather more wood for the fire. The women move into the shelter and begin a soft chant to give guidance to the travelers.

The fire spreads warmth throughout the shelter. The men also gather around to discuss the results of the hunt.

Boa notices Bo-nah's wounded arm, and the hide and poultice his daughter has applied to it. He points at the wound with a questioning look. Bo-nah' relates the story by gesture, and by showing Boa the sites of various blows and wounds inflicted on the animal that now lays on the nearby palette. He seeks permission to lay down to rest as he has pushed himself very hard to help bring the animal and their weapons back to the group's sleeping place. He especially praises Ah'-oom's leadership on the hunt. Ar-gah' shows Boa the leg bone that Bo-nah' had prepared. He picks up a spear from the palette and fits it to the niche. He demonstrates how to hold the two pieces but does not throw the spear. Bo-nee' has sat down, but he is able to express his great pleasure at the skill that Ar-gah' has mastered.

Ar-gah' indicates the two travelers who have remained at the edge of the firelight. He describes the experience he and Ah'-oom had in their meeting with the strangers. He tells them both that Ah'-oom had stayed with them and would be leading them to the sleeping place. Boa looks concerned, but Ar-gah' relays that he and Ah'-oom shared food with them. He believes they are a group travelers who have their own food and shelter and will cause no harm. He describes their absence of body hair— more like Ah'-oom; less like he, Boa, and other members of their hunting group. He also tries to describe their peculiar way of talking, more like singing than the barks and guttural vocalizations of their own group.

When Ar-gah' tries to emulate the singsong voice of the travelers, Boa and Bo-nee' bark in laughter.

Now, Ar-gah' motions the travelers forward. Each lays his spear down, then approaches Boa and extends an open hand in greeting, both touching Boa's upper right arm. Boa responds with a low grunt. He bids the strangers sit with him. He motions toward Yah and Yah'-nay that they be fed.

Rain begins to fall. The roof of the shelter starts to leak, and the men move hides around in the hopes of keeping the inside of the shelter as dry as possible. The fire in front of the shelter is built up strong enough, it is hoped, to persist even through a heavy rain. The women continue their chant in the hopes that Ah'-oom and the travelers will hear their call. The white-feathered raven flies from his perch on the shelter roof toward the north, in the direction of the returning group. Yah'-nay continues her soft chanting as she watches the bird go.

Ar-gah' motions that he will keep the fires going. After a while most of the members of the group fall into a deep sleep.

Later, while scanning the dark beyond the firelight, Ar-gah' spots Ah'- oom walking alone along a path to the sleeping place. He gets up and moves quickly to her, being careful to avoid the broken ground that could cause serious injury. Ah'-oom indicates that the travelers are coming, but that their burdens had grown heavy. Ar-gah' and Ah'-oom go back to help them. Once they take up the loads, the two hunters and the travelers make their way to the fire and shelter.

Ar-gah' goes to Boa and Bo-nee' to tell them of Ah'-oom's return with the travelers. Ah'-oom wakes Yah, Yah'-nay, Pe'-dee, and Pe'-dah. Once awake, the women once again ululate with joy at the safe return of all the hunters and the blessing of the meat. Boa and Bo-nee' hug the returning hunters and bid them sit down by the fire.

The women who have been chanting grow quiet now. Yah'-nay approaches Boa and gives him the beautiful blue stone, which she has woven onto a colored hide backing. Boa is momentarily surprised. Then, he calls Ar-gah' and Ah'-oom to his side. He looks into their eyes for a few moments, before putting the decoration on Ah'-oom's chest. She is overwhelmed. She bows her head then hugs her father. She turns to look at Ar-gah'. He has no expression in his downcast eyes.

Ah'-oom frowns. She once thought Ar-gah' impulsive, a showoff, a weak spot in their hunting party. Now, she sees him as valuable, as a friend. She steps toward him and puts the honored decoration she has received from her father on Ar-gah''s chest. His expression changes to one of joy and pride. He tries very hard to suppress his emotion.

Bo-nah', who has woken up, looks at Boa and bows his head toward his older brother to indicate his approval of all that had occurred. He looks at the travelers now joining his family around the fire. He knows they are from the north, but he wonders where. He begins to imagine places there, filled with many wonders. He resolves to go there someday, perhaps with the travelers when they return to their homeland.

The black sky along the eastern horizon turns gray with the emerging sun.

END

Commerce

Late summer, 3000 BCE. Memphis is a trading center between the southern valley and the northern delta of Egypt's Nile River, crossed by the trade route connecting western North Africa and Mesopotamia. Hesina, daughter of a successful commercial warehouseman, awaits a caravan arriving from an oasis in the western desert. The caravan carries leather goods, dates, fruits, and other goods

====

Late summer on the Nile. Its source high in the mountains that lie southwest of the Horn of Africa. The lake, covering 26,000 square miles, would become known in modern times as Lake Victoria. From this source, the longest river in the world flows over 4,100 miles through deserts, mountains, and multiple outlets in its delta to the southern shore of the Mediterranean Sea.

Now, in the time of this story, the Nile has returned to its banks following the annual spring flood, once again restoring fertility to the vast territory in northeast Africa that will eventually become known as Egypt.

In the city of Memphis, Hesina is a young girl, intelligent, a quick learner, and eager to find new things to learn. She helps her father, Senna, with his warehouse business, linking the traders along the Nile with traders from the eastern and western deserts. She strives for improvement in these businesses wherever and whenever she can. Such are Hesina's talents and intelligence that several tutors have been engaged at the suggestion of Pharaoh Djer, grandson of Narmer Menes and friend of Hesina's father from an early age. Because their patriarch is the man credited with joining Upper and Lower Egypt into a single governing entity, the power of the family of Narmer Menes is unquestioned.

One of Djer's recommended tutors, Amenemhet, introduces Hesina to a woman of nearly the same age, Nena, a member of the royal household who is also under his tutelage. The two young women have become close friends over the years.

Here, Hesina begins her story.

Part 1.

I am sitting cross-legged, sipping hot tea on the roof of my father's warehouse as dawn breaks over the western suburbs of Memphis. The early morning light casts a weak shadow of my body as it falls across the roof in front of me. I am looking away from the dawn, though, staring intently into the still-dark west as I wait for Alim's caravan. He will bring us baskets loaded with dates, nuts, and precious but perishable figs from an oasis two days journey to the west. The thought makes my mouth water.

I had been alerted to Alim's arrival by his son, Rabiah, who serves his father by acting as runner and messenger in advance of his caravan of burros, camels, and other beasts of burden, all carrying items for trade.

I consider our options in light of Alim's approach. If my father and I can trade with him for the figs, we will have to rush to distribute them to our vendors and most favored customers before they spoil. I tap the cup thoughtfully against my lip. Among his less perishable products Alim may be trying to trade dates. If Father decides to buy these less perishable products, we will have to find a company to ship the dates north along the Nile, through the delta, to the Mediterranean port of Rhakotis, and from there by sea north to the ancient city of Jerusalem. There, the dates will be accepted and passed along to other traders and merchants by our longtime agent and business partner, Laret.

I grin as I stand. I'm not sure what I'm worried about. Alim is a tough negotiator, but he was not happy the last time he did business with us.

Father had recognized my talent with numbers and hieroglyphs early on, and had enhanced my natural talents by having me trained as a scribe by the priests in the temple of Ptah. My skill with the new ways of tallying sums using pebbles and grooves in the ground had gotten Alim to agree to compensation that Father knew gave him a rather slim profit margin.

As I descend into the warehouse, I wrap my shawl more closely around me. Mornings are chilly here in the desert, though I'll likely wear nothing more than a cotton wrap once the heat of the day begins to simmer.

Despite my prowess in our last negotiation with Alim, I know that the nuances of my father and the western trader's relationship are valuable, and therefore at least as important to any trade negotiation as my skill in recording our wares. Even so, I will focus on the part of the transaction in which I excel. I've moved onward and upward, from pebbles and grooves to the permanence of clay tablets.

Keeping a record of trades by making impressions in wet clay has been in use for some time in the distant cities along the Euphrates River. Or so I am told. I learned of this and of the clay tablets from a trader who recently brought jewelry from the city of Uruk far to the east. The trader had come to Memphis to trade his jewelry for a quantity of cotton grown along the Nile, the kind used to produce cloth by our weavers here in Memphis.

Once the jewelry trader was able to complete a deal with Father for a large quantity of cotton, I watched as he pressed his codes into the soft clay of the tablet. Once dried in the sun, the trade information could be stored permanently, even longer than that recorded on papyrus. As the trader bent to the tablet his coat fell open to display a neckpiece with a hanging pendant made of an inset jewel of deep blue streaked with gold.

He explained that the stone had been passed down among his people. It was an honor for him to wear it. In my eyes, the stone was as beautiful a decoration as I had ever seen.

My father has encouraged me in all my explorations and experiments with these new ways of recording values and doing trades. He also enjoys exploring new ideas for his several trading groups with me, and I know he hopes that I will follow in his footsteps many years from now.

By the time of the journey I wish to share, Father had already built a substantial trading and warehouse business by working with traders and shippers who needed storage space.

As such, he had turned his sights to a new venture in boat building and shipping. Father has set up a workshop in this complex of buildings and has engaged the services of a man who has designed some of the successful boats, which are capable of carrying grain and livestock to us here in Memphis from the irrigated agricultural areas along the upper Nile.

Because he has already established warehouses for grain in our complex, Father will not have to pay for storage any time he negotiates with these grain buyers. A savvy business decision on his part, I'd say.

I have a lot on my schedule today, but I make a note to talk to Father about the trader's beautiful blue and gold pendant. I cannot keep it from my mind. Father agrees that the blue stone might be valuable.

====

"Building a boat capable of commercial travel on both the Upper Nile and the Nile Delta is foolish," I say. I can't help my truculent tone or the way my arms have folded themselves across my body. I stand in a room my father reserves for himself within one of the warehouses. Father has told me about his idea for a vessel to do such a thing, but the numbers in my head just don't add up. "The shipping from growing areas in the delta is too scattered and undeveloped, and the channels are too variable year to year. Besides, boat built for shipping in the delta would never be fast enough or have enough capacity to compete with the bigger, faster boats built for shipping bulk grain. "Far better to build two boats; one for each type of shipping." Hesina paused. "But it is even possible that any boat built for the delta cannot earn enough to fatten the single owner no matter how well designed, and no matter how fit and eager the crew. Some kind of association of shippers may be feasible, but it will take time to find enough warehousemen and shippers with enough uncommitted resources to make it work long enough to establish a value for the service."

Father inclines his head in my direction and chuckles. "As always, daughter, your insights into the problems of business and trade opportunities are as good as any in Memphis. I believe your thinking is correct, but I still wish to ponder the idea of building a single boat to travel both waterways. I like the idea of forming an association of shippers, though."

He drums his fingers on the wall. "Once we establish the delta service from Memphis, we can think about a warehouse and port on the Mediterranean Sea. But that is in the future. For now, my daughter, I have another business proposition for you to consider. I have been aware of the jeweler from Uruk, and his flaunting of the blue stone among those with refined tastes and plenty of money. He has been here before. He spreads precious fragments around like a monkey scattering his turds, and the activity is having an effect. Many are talking about it. If I am able to find a supplier, and a foothold in the market for it, I think we will gain much if we can bring some quantity to Memphis."

"You know my interest in expanding our enterprise, Father," I say, "but you have not yet tried to expand into markets with sources of supply beyond the reach and protection of Pharaoh. How will you protect your Mesopotamian suppliers when they are trying to bring a shipment of blue stone to you here in Memphis? I don't think your experience so far with your sales agent in Jerusalem will be of much help. Besides, I am not sure I trust Laret. The quantities in his reports do not always add up, and the mistakes seem always to go to his advantage."

"I am well aware of Laret, and his loose connection with the reporting of his trades. To some extent I forgive him because his connection with hieroglyphics is also loose, and with hieratics almost impossible. Worse, he does not have much experience with the counting systems you have set up. But he performs other services for us that I do trust and cannot do without as long as I seek to trade our goods anywhere in Jericho, or anywhere along the Jordan River to the Sea of Galilee. As I have told you many times, I dream of expanding our enterprises into the lower Euphrates River in Mesopotamia, and Laret and Jericho are my stepping stones toward that goal."

"Yes, Father," I reply. After a pause to drink some beer, Father continues. "At any rate, the Uruk jeweler and trader in Lapis is named Faiza. He will be returning to Uruk soon, and he has invited me to send one of my personal representatives with him to explore whatever markets we might have an interest in.

"I want you to be that representative, but you and I know that the venture will be filled with great danger and great discomfort. The rewards, if any, could be as smoke. They could blow away in the first breeze. Please give this some thought and speak to me about it in a month.

He has come to his point. I know from the way he massages his left ear, a telltale sign that I should listen. "I know that I will feel more confident in you traveling such a great distance if you take on a shorter journey in the meantime," he says. "So, I want you to go upriver with Serenen to check on the irrigation project that our friend, Heti, is working on near Beni Hasan. He is trying to increase the quantity of grain that he can plant and ship to us in the spring. I need your assurance that he is staying on schedule. If he is having trouble, I need to know what is causing it, and I need your ideas for resolution of any difficulties you might find. You will leave tomorrow.

"I know this is a lot to ask, my daughter, but I have many things going on, and I need your help in making sure everything is going smoothly."

My heart beats quickly under the hand I have brought to my chest. I am certainly excited by my father's request, the responsibility he is lavishing upon me is tremendous, but I remain terrified by all I do not know. What if I fail him? "But Father, I have never been upriver," I protest. "How will I know what I am seeing?"

Father still seems confident, which buoys my spirits. "Serenen has been upriver many times, and he knows Heti very well. They worked together to prepare the fields that now produce so much good barley and beer for us. I want you to have the experience of traveling with Serenen because I think he will need to be one of your companions on your trip to the north with the Mesopotamian trader... if you decide to go, that is."

I know that he expects me to go with the Mesopotamian jeweler, and that he will be greatly disappointed if I choose not to. The dangers, discomforts, and difficulties of the trip give me a great deal of concern, I must admit. However, the opportunity for adventure and learning and exploring new possibilities for enterprise mean that I have very little reason to say no.

"Yes, Father," I decide. "I will tell Serenen to prepare so that we can be ready to go upriver in the morning." Father clasps me up in a hug. "Excellent. Be sure to talk to him about the trip to Mesopotamia, too."

"Yes, Father," I reply.

***

I couldn't sleep that night. Lying on my cot on our roof, I tossed and turned. One moment, I would think of how excited I was to feel the breeze come off the river as our boat was steadily rowed up the Nile. I could see myself ducking out of the way as the sail setters moved around the boat trying to catch the nuance of each breeze. The next, my heart would plummet into my stomach as I prayed to the gods that I wouldn't disappoint my father.

Early the next morning the sun rises on a stiff northerly breeze. I rise, knowing that the breeze will mean a faster trip up the river to Beni Hasan—possibly five days instead of six.

One of the Nile's many boatmen, Akar, a man well known to my father and me, is sailing upriver with his crew of ten oarsmen to Thebes. At Father's suggestion Akar has invited me to come along with whatever staff and equipment I need in the irrigation fields. He will drop us off at Heti's, then turn around to pick us up on his way back down the river.

Akar has now docked close by our warehouse. The big, muscled man moves toward me and greets me with a familiar, teasing tone. "The boat is ready, but first, my sweet beauty, you must serve me tea."

"Of course, Akar," I reply with a roll of my eyes, though all of my teeth show in my smile. "Do you think you will have any problems finding your way upriver? You will be hauling precious cargo, as you know." I let a smirk cross my face, then motion to one of our household slaves to bring some tea and honey-sweetened biscuits. "It looks windy this morning.

Can you tell me how the trip will go?"

"It may be a bit rough because of the wind, but the disturbances from the summer flooding have now subsided. I have a very capable crew. With our sails out we should be able to capture most of the north wind. If we do a good job of sailing and rowing, we might be able to cut the time to less than five days." I'm impressed to hear this. "I'm sure you know, but this will be my first time on an extended trip upriver. Are there any dangers that I may not be aware of, that I might have to prepare my staff for?"

"There is a lot of traffic moving up the river with tools for the irrigation and planting, as are you and Serenen. Have you made arrangements for pick up and return to Memphis "Yes. We are allowing ourselves five days helping clear the irrigation channels at our fields near Beni Hasan, then my father has made arrangements with our friend Kames to bring us back. Do you think Kames is a capable boatman?"

A raven with a single white feather lands near us as we talk. I toss it a morsel of biscuit, and the bird hunkers down as if to give his thanks before grabbing it with his beak. I smile and turn back to Akar, who seems to be considering my question. "Well? Is he a good boatman or not?"

Akar sucks his teeth. "He is a good boatman if he can find a good crew. He usually can, but that is never guaranteed coming from Thebes. Your man, Serenen, is a good one with either an oar or a sail. I think he is fully capable of running a boat of his own if you and your father ever decide to put your own boat in the water.

"In any case, on your return to Memphis you will be traveling downriver with the current. That is a much easier task in terms of effort and navigation, but never forget how dangerous the Nile can be when you are coming downriver. The river is vast and relentless. She doesn't care a fig about the humans who try to earn a living on her. Remember: things happen fast when you are in control of your boat. They happen much faster when you have lost control of it while going downriver on the Nile."

By this time, Serenen has loaded the tools, equipment, and two weeks of foodstuffs onto the boat with the help of Akar's crew.

It is time to go, but I am not yet ready. I stand and stretch lazily. "My father is sending one more passenger to go with us. We need to wait for him. I also sent word to the Pharaoh's palace this morning, asking my friend Nena to travel with us to keep me company. I hope this will not add too much of a burden to you and your crew, Akar."

"Of course not, my pretty one. I know your friend Nena. She is a very charming woman, and I know her to be among the circle of friends who are familiar with Pharaoh. You can count on my complete discretion.

Who is the other passenger?" "I believe the one coming up the street is our last passenger." A man comes up the path. He is short in stature but well muscled and at ease on his feet even though his burden appears to be heavy. "This is Paru, a lieutenant of guards in the palace of the Pharaoh. My father has asked Pharaoh Djer to send him with us to protect us, and to assure our peaceful intent to any on the river who might inquire as to the Pharaoh's interest in our trip."

"Do you feel this extra protection is necessary? Do you and your father not trust us enough?" asks Akar. "Please put those concerns away, Akar. They are not worthy of you. My father wants Pharaoh to be fully aware of our plans for the property at Beni Hasan, and he has sent his most trusted and most physically capable lieutenant of guards. Once we are underway, Paru can assist with the rowing and sail setting. He is under your command for anything to do with the operation of the boat on the water, even if others along the river have hostile intent toward us.

"If a minor problem escalates, though, I will want you to defer to Paru's experience in combat. If he calls on you or on any of your crew for assistance while we are under threat, I will expect you to give him all that he may need." I nod toward the newcomer's satchel. "You have seen the heavy bag he has brought on board. I can assure you that the bag contains weapons and that Paru is very well experienced with their use."

Akar begins to protest, but I hold up a hand. "To be clear, Paru is here primarily as a diplomat, a representative of Pharaoh. I do not expect trouble on this trip, but if trouble occurs, Paru will be in charge of our defense."

Our captain inhales, and as he exhales, his fighting spirit seems to deflate.

"Yes, Hesina. You have my full support in this."

"Thank you, Akar." I see a friendly face coming up the path and brush the crumbs from my skirt. "Ah! Here comes Nena. Once she is aboard with her things, we will be ready to sail."

***

Nena is a tall and darkly attractive beauty, though, today, she is dressed modestly with a shawl to protect herself from the wind.

I rush off to greet my friend.

"Hesina, hello," she drawls, draping me in a hug. "I wanted to bring more bags with me," she says, indicating the servant struggling under two trunks, "but I was told there wouldn't be room on board."

"Hello to you, too." I hug her back and turn to take in Captain Akar's boat. "Well, what do you think?"

Nena does her best not to simper. I can tell she's thinking about all of the barques she's floated on, accompanying the Pharaoh and his family on their river cruises. "Charming," she settles on "Quaint."

I don't let her comments deter me from taking her hand and guiding her on board. "This is our own cabin," I say as I show her the area Akar has draped off with patterned cloth for our privacy. I almost do a little skip step. At home, I sleep on the roof with my parents. This is the first time in my life I've ever had my own private cabin.

Nena peers around the corners. "Where are the beds?" she asks. I shake my head and remember her telling me about the woven cots that litter the Pharaoh's palace.

As we return to the main deck, I hear some of the off-duty rowers tuning instruments. Soon, drums, tambourines, and lutes regale us. I know that Captain Akar wants to be as professional as possible—he is on business for my father, and I would do well to remember that I am, too—but I see that even he is tapping a toe to the music.

The rest of the crew begins to row, muscles rippling in the sun, and we are underway. A deckhand gestures to a table full of food and encourages us to partake. Small beer, bread, dried raisins and dates, and some salted meat are laid out on the wooden table. I take a bite of meat, savoring the way its saltiness melds with the sweetness of the raisins, and I rinse my fingers in the water bowl.

Nena looks up at another member of the palace entourage. He has brought aboard a small pot of expensive honey, which he is slathering on a piece of bread. "Djal," she hisses and subtly flicks her wrist as though to ask for the honey pot.

Finally, I chuckle. I throw an arm around my friend's shoulders. "I know this is all rather hard living after the palace and our day trips to see the sites around Memphis. But this trip upriver is an adventure; ...a real adventure."

I can't stop the smile from crossing my face. My hair whips in the wind as we head up the Nile and closer to Beni Hasan. Nena seems unsure.

***

Once darkness falls, the workday for the boat crew is over. The dangers of trying to travel on the powerful river when there is only moonlight and it is too dark to see are too great for any but the most skilled crew motivated by the greatest of possible emergencies. Toward the end of the day, Akar looks along the shoreline for a place to park the boat where we will be safe from thieves, crocodiles, and hippos, which might have too great an interest in the human contents of the boat.

I listen to Akar instruct his men on the procedure for their night watch.

Each of the rowers and sailors has an assigned time when they will have to keep watch for danger. A dripping pot of water marks their time on guard. When it is empty, the watcher will rouse the next watcher out of his restful sleep. Any watcher who fails to stay awake has been threatened to expect several lashes from Akar's whip.

"I'm grateful to not be a watcher," I whisper to Nena.

She yawns as she nods her agreement, and we settle into our cabin for the night.

***

The rest of our journey and our arrival at our fields near Beni Hasan has been uneventful. We meet my father's associate Heti and his family and set to work clearing the irrigation ditches.

Akar the boatman continues upriver after dropping our party and our tools and supplies off at our fields. He thought that he would not return downriver for many days but told me that he would send a message to Kames; that he should be prepared to take us back to Memphis within the next several days. "However," he had said, "if I cannot find Kames, or if he is not ready, I will come back downriver myself to pick you all up for the return to Memphis. I will plan on seeing you again in five days if that becomes necessary."

I have always known how to work. Without the benefit of sons to pass the business on to, Father has not kept me from the necessary work of his warehouse. Now, after several days in Heti's fields, I know the sore muscles and blistered satisfaction that come from turning the fertile ground to produce the barley and beer that is the backbone of our trading business. Heti's fields do not seem to suffer from lack of water or from too much water. He and his family are grateful for our help, though, as there is much to be done.

We work all day and much of the night while in our fields. At first, I feel shy to work without a wrap in the heat of the day but toiling with sweat and the risk of heat exhaustion quickly rid me of that. By the end of the second day here, I came to think nothing of stripping off my clothes as the sun rises higher in the sky.

The work is hard and makes the small pleasures of life all the more valuable. We eat and laugh together at night, and we sleep well under the stars.

At first, Nena wasn't sure about working in the fields. She worried, and with good reason, about bringing embarrassment to the house of Pharaoh as a woman of her class tilling the fields and being seen naked in them.

However, the work was too much for our small group to handle. On the afternoon of the second day, in a fit of pique brought on by Nena wondering why I hadn't removed my wrap, I snapped at my friend to get off her high horse and dig. She may have been upset that day, but soon, she also got into the habit of working, bare-skinned, in the hot sun. I know my friend well enough to know how good it feels for us to pull our own weight out of sight of our parents and overseers. Now for possibly the first time ever we feel worthy of the food we put in our mouths.

There is a niggling worry in my mind that I have neither seen nor heard from Akar or Kames. On the sixth day at Beni Hasan, we are clearing a particularly difficult drainage ditch. My hope is that dislodging the bracken that's built up within it will allow the river to flow into this patch of crop.

Along with her clothes, Nena has dispensed with her ladylike decorum.

"Hesina," she asks, raking the silt, "what do you call a water buffalo with a human face?"

I pretend to think on the answer to her joke, which is sure to be off-color, but I don't get to hear the punch line. I look up from my work to see Akar's boat on the river, coming toward us.

"I'm not sure, Nena," I say, "but would you run tell Heti's wife to prepare eleven more settings for dinner this evening? We have company."

A flash of the old Nena is back, as she primps her hair and dons her waistcloth before running to the farmer's house. Heti was delighted that such beautiful women would help in the fields and show off their womanly charms at the same time. He, his wife, and another older couple lived on the farm to do the planting and help fend off the creatures who would come to feed on the green shoots when they popped out of the ground.

The other members of Heti's family, and their two friends also worked naked, but they looked away from the young women. They all had a note of sadness in their eyes, perhaps remembering their own younger days.

I lay down my rake and rush to greet my friend. As I come closer, I overhear Akar instructing his oarsmen. "Tie the boat up near the trees, then stay out of the heat for the afternoon. By late afternoon, when it is cooler, bring the boat back up the river to these fields. Maybe we will get a good meal tonight."

As his boat departs, Akar approaches me. He must be amused to see a merchant's daughter naked under the sun like a commoner, as he says, "Well, my beauty. Hard manual labor becomes you."

"Be careful of your language, Akar," I warn. "Do not take advantage of my need for comfort in this heat. How was your trip upriver? I am very glad to see you back here by the way." "The trip was good, but Kames was nowhere to be found and my own plans were thwarted when the men I intended to meet with could not be found either. I will be delighted to take you and your party back downriver, but I need to know how much more time you need in the fields."

I wipe my brow and survey the field as I consider my answer. "If Nena and I and the families here are the only ones working, then it will take us another full day. On the other hand, if you can release your boat crew to us, we can probably finish in half a day."

Akar nods. "Aye. They shouldn't be too sore from our light row this morning." Akar paused and looked down at his feet for a moment.

"I will make you this offer, pretty one. If you and Nena will put your clothes on while my men are working with you, we can probably get all the work done. Do you understand me or do I need to explain my concerns further.?"

"Yes I do understand, Akar. I understand completely. Nena nodded in support. We will cover our bodies when your men are around. "Wonderful." I grin. It feels good to be the one delegating for once.

"Now for more important things. I have alerted Heti to your arrival and have asked him to prepare food for this evening's dinner. In our time toiling in these fields, we have discovered that the palace guard Paru is a poet of considerable ability. If those of your men who can play instruments will play for us tonight, it will be a joyous event."

After Akar has called his men back to Heti's dock, we all head for the main house, Paru greets our shared friend. "Did you have a good trip upriver, captain? See any crocodiles?"

"No. No crocodiles," Akar answers. "A few hippopotamus only. They stayed well clear of the boat."

"Good. I would have hated to not be on board to wrestle a crocodile for you."

Akar tips back his head and laughs. "Hesina tells me you are concerned about our return to Memphis. Something about nudity."

"Yes. Something about nudity. It is a custom in the rural areas for Egyptian farmers to work without clothes in the heat of the sun. Hesina and Nena have also picked up that habit and seem to enjoy the feeling of freedom. The men on my boat also work without clothes when the sun is hot. My concern is that the nudity is bad when there are two such attractive women on the boat, who also enjoy nudity, and seem not to worry about the effect of it on men. "As captain, any problems that might arise from the temptations that could overtake some of the men, could put the boat itself in jeopardy.

“I want to require that everyone cover their nudity on the boat while we are underway, no matter how hot it might get. That will include you, Paru, and Serenen, and it will especially include Nena and Hesina. Hesina and Nena have both agreed but I need the support of you two men if I am to enforce the rule.

“I believe it will be very difficult to enforce the rule among at least a couple of my rowers. Since I need to be in full control of the boat and the rowing at all times when we are running downriver, I will have to have one or both of you ready to apply whatever punishment might be needed to calm any bad situations. Do I have your agreement on that?” Akar asks.

“Yes. Of course. “But you know your men, and you know the potential trouble-makers. How will you communicate your concern to us, and how will you show us the problem so that we may act on it?” Paru asks.

“I will tell my crew what I want them to do to cover their nakedness, and I will tell them what I have in mind for punishment if any are unable to control their passions... or their erections. They know me to have a strong whip arm, and they know that if I warn them about expected behavior, that I will use the whip on violators.

“How about you, Serenen? You have not yet spoken in my presence. What are your thoughts?”

“Should a problem arise, Captain, I will defer to Paru, unless the problem is urgent and I am closer to it. If I must deal effectively and quickly with the problem, and Paru is not able to assist, I may heave the offender overboard, and take over his responsibilities on the oar, or rudder, or sail. I am well-experienced in all aspects of boat operations.

"Please understand, Captain, that I have been attendant, protector, and mentor to Hesina since she was a small child. I take my responsibilities very seriously."

Paru spoke up. "Be assured of my full confidence in the words of Serenen, Captain. Before he became protector to Hesina he served with me in the Pharaoh's personal guard unit in the palace. We have been involved in many battles. I trust him with my life."

Akar sat back and rubbed his chin as he thought about the words of the two naked men standing in front of him.

Akar speaks. "I am glad to have your understanding and support. I thank you. In the morning while everyone else is working in the fields, I will want to work with Paru to correct any possible sources of problems on the boat itself. Though I will be moving the boat toward Memphis with haste, there will be several long hot days on the river. The women's shed will need to be completely shielded from casual views by the boatmen.

At the same time, to avoid boredom, the women will need to have an opening toward the shore for their viewing, and there must be an opening for breeze. I will assign the boatmen in such a way that they won't be constantly thinking of what the women in the shed might be up to; ...but there will be leisure times when we are under sail, and they might all be unoccupied, and left to their own imaginings. That is the time we must be most alert. Two of my boatmen are particularly trustworthy in enforcing my wishes among the others. I will ask your advice, Paru, on how they should be deployed, and how advised as to their responsibilities.

"Let us now see what we can do to finish this day's work in the fields, and then prepare for the evening meal," Akar concluded.

***

Our party was quiet and didn't last much beyond the last bowl of honey-sweetened dates. All the participants were tired after several days of hard work in the fields and on the boat, and another hard day of work would begin again the next morning.

We finished irrigating the fields as the sun passed its highest point in the sky. After a lunch of bread and a small ration of beer, we were able to finish loading the boat with tools, weapons, and foodstuffs. Akar wanted to get as much time as possible to move down the river before dark, so he pushed everyone to move quickly to finish the loading. I overheard one of the rowers tell him that he enjoyed the life around Beni Hasan and wanted to stay a while longer. Though this visibly irked Akar, the man was not contracted to him beyond this voyage, and we pushed off without him.

"Ah, well," Akar said. "We can make do with one less rower."

Now, we have cast off from the small dock, and into the Nile's steady, inexorable push downriver toward Memphis and the great sea beyond.

Once on the main channel, our oarsmen begin their routine of coordinated rowing. I've learned that it's designed to give Akar maximum control over his boat as he searches the river's surface for the fastest currents.

The rudder men and sailors are tuned to his every hand gesture and to the nuance of his every verbal command. As well-experienced boatmen, Paru and Serenen are alert to the same signs.

Nena and I start the day watching them row, but the repetitive motion grows boring over time. The view on the banks of the river isn't that interesting either, now that I've seen it once. I wish I could be more useful to Akar, pull my weight literally at the oar or clean the deck for the men, but he tells me not to worry myself over it. Eventually, I lie down, somewhat sullenly, and await nightfall.

***

At the end of the day, Akar compliments all on the crew for their good, strong work through the day. As we tie up for the night, Akar tells his men that they have to sleep ashore in the trees, the better to watch for the various creatures, or persons, that might otherwise threaten the boat and our precious cargo.

"I regret that I cannot offer beer to you this evening because there is none on the boat," he says to some grumbling. "You will have to quench your thirst with river water, but I am told that this place has a source of very fine water, and I am sure you will be most pleased. There are extra rations of dates, salted meat, and honey to repay you for your work today.

I'm hopeful that the weather and wind will hold, and we will have a good day tomorrow and the next day."

The captain, Paru, and Serenen lay out their sleeping mats around our shelter and ask us how our day has gone.

I look at Nena and shrug. "It was fine, but we wondered if, tomorrow, we might spend more time outside of our cabin... maybe even take a turn rowing?" I see Akar's expression sour, so I hasten to add "Or set the sail, or the rudder."

Nena agrees. "If we spend another day in pampered indolence, I think we will grow fat, and our moods will be unattractive."

"I am less worried about unattractive moods than you may be," Akar says, "yet I am mindful of your concerns and will give them some thought. I'll let you know in the morning. Now it is time to sleep. Rest well in the encircling arms of Horus."

***

Before dawn the next morning, Akar, Serenen, and Paru come to visit us. Akar has given our request some thought, and Nena and I will be assisting the rudder men, allowing them to take occasional breaks.

The day's sailing downriver proceeds well. Nena and I work hard in our new roles. The rudder men gave us liberal tips on how we can improve our mastery of the procedures involved in changing rudder settings.

By the end of the day, Nena and I were very tired, but also very happy that we had been useful and had performed well. I saw Akar laugh a few times as he captained his crew today, but I can tell he still feels reserved over whether we'll have smooth sailing over our remaining three days of travel. We spent the evening hours building more spears to add to our weaponry. Even Nena and I have our own small stockpiles should we be attacked. As I ready for bed, I overhear Akar tell a mate that before we will pull up to my father's dock in Memphis, there are pockets of crocodiles and hippos in some of the sloughs ahead, and worse—though he has never had to deal with them—there is a nest of thieves at work in this part of the river.

I sink onto my pallet with my stomach churning. "What's got your goat?" Nena asks.

I can't help thinking of a goat, bloodied in the mouth of a crocodile.

"Nena, have you ever seen a crocodile?" She's spent more time on barges than I have, so I know she won't sugarcoat things for me.

"I have, but if we don't go overboard, we won't have anything to worry about. They cannot easily crawl onto the boat."

This makes me feel safer, though I resolve to move more carefully about the deck tomorrow.

***

Making and loading the weapons into Akar's hiding places took some time the following day, so we didn't get underway until the sun had reached its peak in the sky. The rest of the day was mostly uneventful. The two exceptions occurred because the two women were now moving freely from rowing position to rudder position, to sail setting position through the day. Akar had not relaxed his modesty rule, but the sight of two such beautiful women moving quickly among the men, and occasionally bumping into them when the deck shifted required more discretion than two of the boatmen were capable of maintaining.

Akar applied his whip judiciously, to the men as their erections popped up as a result of inappropriate – and possibly unavoidable – touching.

Once touched by the whip, the shame-faced offender did not allow the problem to occur again.

The next two sailing days – our third and fourth on the river – were also uneventful, but late on the fourth day, today, there is a change in the look and feel of the air as we pass beyond some trees on the west bank of the river.

I can now see a sandstorm building on the western horizon. The churning dust cloud is already far higher than the tallest buildings I have ever seen. I can hear the dull roar of the storm as it builds and moves toward us. I am frightened It seems to unsettle Akar, too. He quickly instructs his boatmen to steer for the western riverbank. He calls to me. "We must tie up and wait for the storm to pass."

I nod and adjust the rudder, but as we move closer to the western bank, Akar points to a calm spot there. "Look to the opening of that slough! Collapse the sails and steer for it!"

I'm not sure why we're changing course so dramatically, but the crew trusts Akar. They seem calm even as they follow his urgent orders, so too do I.

As I hold the rudder steady for the sudden change of course, I hear Paru at my elbow. "The sands will be intense, miss. There is no shelter here on the open water. You and Nena must lay down on the deck out of harm's way."

I give the rudder to one of the more experienced crew and comply with Paru's command to step aside. Better that an expert rudder man moves us forward than a merchant's daughter who might drive us into the bank in a way that damages the ship.

From our place on the deck, I can feel the boat turn to the bank as soon as we enter the slough. We bump hard. I am grateful for my place so close to the splinter-infested planks. Those who were not ready for the rough stop fall to the deck around us. I hear the splash of a man going overboard.

I raise my head. "Is he okay, Akar?" "Do not worry yourself," Akar replies, putting his cloth wrap over his face. "Just stay down. Paru and I will help get the man back into the boat."

He orders the boat secured to the shore and secured against the wind and sand.

"Captain, what of the sand?" a rower asks. "If enough of it gets in the cracks around the hull, we'll sink."

I share a look with Nena. Neither of us had thought of that.

"Once we secure the boat we need to get out of the water," Paru says nearby. "There are probably crocodiles in the swampy area beyond the banks of this slough. If they are nearby, it won't take them long to find us."

In the pause that follows, I grip Nena's hand. "Perhaps we need to move the boat further down the slough now while we have plenty of daylight.

It will be stormy and miserable to work in the middle of a sandstorm, but that's better than fighting with crocs while we try to move the boat in the morning." "Good thinking," Akar says. "I'll talk to the boatmen. We will need to post guards with spears to watch for crocs, and we will need to distribute our digging tools to the strongest diggers. The women can keep watch from the boat. They should be prepared to climb the masts from time to time so they can get a better view."

Nena sits up on her elbows, blowing a strand of hair out of her face.

"That won't be a problem, I assure you."

After some time, tools and spears have been distributed, and the men bend to the task of widening the banks. Nena and I watch for crocodiles.

Four of the men accompany Akar and Paru downstream in the slough to clear any underwater mounds of sand that might hinder the boat's passage.

Meanwhile, Nena and I use ropes to rig a way to climb quickly up and down the boat's mast, and I take the first watch near the top of the mast.

Nena moves to the back of the boat to watch for possible threats that might be coming from the river. From this high up, I can see the river in either direction, the sandstorm fast approaching, and little settlements beyond, on the peaceful bank. I think of my beloved perch on the roof of Father's warehouse, and I am comforted, even in these stressful and unfamiliar circumstances.

The horizon looks clear of crocodiles and other immediate threats, but the storm is coming closer. I cannot see beyond the storm front. I will have to climb down from the mast when the storm gets too close. I have no idea what the force of the winds and the blasts of sand might do.

Before I have the chance to find out, Akar comes to the bottom of the mast. "Come down from there, my pretty one. I need to see what the storm is doing."

The wind picks up as Akar climbs. As I watch him go, I can hear it howling in my ears, a low moan like one from the world beyond. The sky on the west is completely dark. My heart pounds.

"Adjo! Djal!" He calls his two most trusted boatmen to his side. "We have time to move the boat just a little before the storm hits," I hear him say. There are words the wind steals from me, and then he is yelling, "Hurry!"

They comply.

"Hesina." He turns his attention to me as he descends. "Please. Get into your secure space. I cannot protect both you and the boat from the storm." Nena, ever a daredevil, decides to climb the mast. She shouts down to me that the storm is very close now. I only leave our cabin to alert Akar.

The men stop their digging and focus on securing the boat to the shore. I am pleased to see that, by their digging, they have advanced the boat to a place within sight of the slough's opening back to the main river.

***

In the end, we escaped the worst of the storm. We lost one sailor overboard then quickly pulled him back aboard. There was wind and blowing sand through much of the night, but no damage to the boat—though quantities of sand had to be moved overboard. Because of the possibility of crocodiles, Akar ordered everybody to sleep on the boat deck while at least two men kept watch through the night. We covered our faces with our wraps and waistcloths to keep the sand out of our nostrils and mouths as best we could.

As soon as there is enough morning light to see beyond the boat, Akar climbs to the top of the mast. I watch him as he ascends over the lip of the privacy cloth he had put up for Nena and me to sleep behind on our very first outing. He looks around the boat and reports down that he does not see any threats. "We need to get everybody up to begin digging so we can get back to the main channel as soon as possible. With a good day's rowing, we should be within sight of Memphis by dark."

Memphis, my heart murmurs. Home.

Akar catches sight of me and calls to me as he climbs down, disappearing from view behind the curtain. "Hesina, I hope I'm not prevailing on you and your friend too much, but with two men down, I need your help rowing."

My heart already has felt too full, and now it is bursting. I am proud to do so, and I look forward to telling my father of my active role in returning us safely home. I nod. "We will do what we can."

Nena is awake, too, and is getting some food before beginning her turn at the top of the mast. Suddenly there is whistling in the air and a loud sharp cry from Nena. I turn from my conversation with Akar to gape at my friend, where she lies collapsed to the deck, with an arrow through her thigh. I cannot help screaming. I stopper my mouth with my hands to stop myself.

Paru, luckily, has a more even head than I. "Thieves!" he hisses. He moves immediately to his cache of weapons and motions for everybody on deck to stay down and get out their own weapons from their hiding places without showing them until he issues his orders. I stare in blind fright at my friend, who lies, biting back a yell, on the deck. This is the worst thing any of us could have imagined. I believe that even Paru did not wish to command our vessel, but now, my friend is pierced by a marauding arrow and we must defend ourselves. Even me. I, who only feel at home with tablets and styli, must now defend my life with a spear.

Paru raises his head up above the lip of the deck until he can see and assess the threat. "A boat," he whispers near me. "Two archers."

"Can we capture it?" Akar asks hoarsely, from Paru's other side. "Take it as a prize?"

"No," Paru says. "We are still stuck, and they are moving too fast down the river." He turns to look over his shoulder and makes a motion to two men. They nod. On his command, they rise with him and loose their arrows toward the hostile boat. All three arrows find their marks. I peek over the lip of the deck. The two archers on the hostile boat have fallen.

None take their place.

Paru and I rush to Nena's side. Serenen brings a knife to cut the arrow away and a cloth to bind her wound and stop the bleeding. Paru is not yet sure whether the arrow has hit one of the vessels in the leg that carry a lot of blood, but he assures me that Serenen is the most experienced healer on the boat. Paru motions to Akar to get the men working on freeing the boat from the silt of the slough; Nena's life may depend on how fast he can get us to Memphis.

The morning has suddenly gone terribly wrong. I fear for my friend, but I remember my promise to Akar. Nena is in good hands with Serenen. Along with all of Akar's guests and crew, I bend my back to the task of freeing the boat and begin rowing as fast as I can to help get us to Memphis before nightfall.

***

Akar pulled his boat up to my father's dock early the next morning.

Paru ran as fast as he could to find Father and to look for a priest or physician to look after Nena's wounds. Despite our best efforts to stop the flow of blood with a mud poultice, I spent a sleepless night concerned that I had not done enough. We tried to keep my friend warm, but she grew pale and struggled to breathe. I feared for her life.

Paru soon returns with my father. I had no idea how much I needed to see his face, and I weep as he hugs me.

"Paru said there was a sandstorm. Were you all right?" His voice is shaky.

"I'm fine. Better now." I give him a great hug in return. My mother arrives on the scene with a hooded wool garment to cover me in place of the thin cotton I had covered myself with after we left the slough. I had not thought much of my nakedness since we pulled into the slough ahead of the sandstorm. I accept my mother's warm covering gratefully. I know that I will need days to try to grasp and absorb all that I have learned on my trip upriver. I know that I have done things that will make my parents proud, but for now, in the shock that follows our violent adventure, I am grateful beyond belief for the love of my parents.

I can hear Father cautioning Paru against calling a priest to look after Nena's wounds. "They are too enamored of their knowledge of useless potions and leach treatments that they think cause healing. Actually," he corrects himself, "they are worse than useless. To the extent that they interfere with a careful analysis of what the actual disease or health condition or wound is, their application can lead to the condition growing worse.

"I believe the best treatment now is to have Serenen and Hesina continue Nena's care here. There is no need to move her further. Nena is a strong woman, and I believe she will recover from this wound. As always Serenen has brought the correct healing practice to her, and I believe she will survive," my father said.

***

Nena's wound did respond to the care shown her by Serenen. After some time her color returned and she became again the adventuresome Nena of our long friendship.

My father was very grateful to his friend the boatman Akar for safely bringing his daughter, her guests and protectors through events of such great danger on the Nile. In an inspired moment he invited Akar and his entire crew to join our family and Nena at dinner a few days later after they had all rested. As he did on the trip Paru came, now in formal uniform representing Pharoah. This way those of us who had been on the great journey to the upper Nile were able to share all of our stories about the trip and would be held to the truth of our tales by the others.

Between stories many gifts were pressed upon Akar and his crew by my father on behalf of our family. For his part, Akar told us a story of his negotiation over a thumb-size piece of beautiful blue/violet, almost translucent stone with a nomadic Nubian trader from the mountains at the source of the White Nile, the major source of the Nile itself.

As guest of honor Akar told his story first. "I had seen the stone in the Theban market while passing by a trader showing the stone to another who was interested in it. When I asked to see the stone the trader, an old, stooped and wrinkled black man with an ancient gray-haired burro by his side was rude, insulting to me and all who served me. He cursed me for my family roots far down the Nile. He said we were unworthy of his attention. He was impossible to deal with according to Akar. But the stone was so beautiful in its' deep blue/violet color that Akar told us he could not turn away from it. All in his audience were fascinated by his story.

"I could see into the stone almost to its heart. As I turned it over in my hands the stone almost came to life," according to Akar. "It felt warm and seemed to grow warmer the longer I held it. I knew I had to have the stone!" Akar said with enthusiasm.

"With a fearsome scowl on his face the old black man finally agreed to sell the stone to me and then quoted an outrageous price for it. I agreed to the price without a moment's thought. His demeanor at the end of our exchange was so fearsome that I thought he might spit at my feet before walking away. He did not spit but, after a few steps, he did wave his arm at me with a rude gesture that any Egyptian will recognize as very offensive.

By now the burro had turned in my direction and began braying in a particular way that was, I'm sure, intended to be even more offensive."

Akar is a story-teller and no stranger to the verbal tricks needed to have his audience on the edge of their seats on the way to the end of his tale.

I was not able to restrain myself. A loud question burst from my mouth.

"Can we see the stone Akar!"

All nodded their heads enthusiastically. My father also added his voice.

"Yes Akar. Do you have the stone with you and can we see it?"

Akar looked at me and then to my father.

"Yes, Senna, my good friend. I will do more than let everyone see the stone. I will make the stone a gift to you and your family. In that way I believe it will be a permanent tribute to our long friendship and success in commerce. Also, it will always be available to your beautiful and very ship-worthy daughter Hesina. She and her friend Nena, equally skilled as a sailor, are welcome as crew on my boat any time," Akar said as he lifted the beautiful stone from his purse to the shouts and gasps and applause of all those gathered before him.

It was a wonderful evening.

I learned from my father, later, that the Uruk trader in those other rare blue/gold stones from the distant lands beyond Mesopotamia had to leave sooner than he had hoped. There was a disruption in deliveries he had a personal interest in because of a shipping problem near the ancient city of Jericho. He begged forgiveness that he had not been able to wait for my return but promised to return to Memphis next year. He hopes that we can renew our plan to take Father's representatives, to Uruk to investigate trading opportunities in the gem trade. My heart was full to bursting because father had already assigned me the task of leading that delegation.

As a token of his commitment, the trader left a sizeable piece of the blue stone with brilliant streaks of gold to tickle our imaginations while we await his return.

Of course we feel that Akar's gift to us is at least as desirable as the blue/ gold stones brought to us by the Uruk trader from the north, but we had no idea where the old Nubian trader had found the stone, nor where to begin a search for him or for the source of the stone itself.

I sit, pondering these stones on the warehouse roof this morning. I look to the east and see a new caravan from the east with new goods and materials that will need to be appraised, purchased and taken into my father's warehouses;...and the traders on the caravans will have stories to tell of their adventures in the desert on their journey here. I fly down the steps to greet them with honeyed dates and portions of beer.

I cannot wait to hear their stories.

END

Empire

Midwinter 330/329 BCE. Aello, a young woman, is a skilled armorer with the army of Alexander the Great of Macedon. Alexander is moving his army of 50,000 soldiers and camp followers into southwestern Afghanistan near present-day Kandahar. Aello sees the rugged peaks of the Hindu Kush Mountains, far to the northeast, shrouded in dark storm clouds.

====

Part 1

"It is an honor, gentlemen, but it will be very difficult," he says, "to leave the community where I grew up, had a family, and grew a thriving business."

I walk into the space we use for our design work and pause as my father continues to discuss the work of our metal shop with two men in battle dress typical of senior field officers. They are representing the man everyone calls Alexander the Great of Macedon. He is the man leading the army that has scared the despised Persians out of our beloved Egypt without a fight. The high priest even went so far as to name him the "Son of the Gods." Now, the word on the street is that Alexander is mounting a march to the east.

My father sees me and beckons. "Aello. Come here please and meet these two gentlemen."

My father indicates the taller and probably more senior officer.

"This is Captain Kallias, a staff officer in Alexander's army, and this is Lieutenant Petros, his associate."

I shake hands with each man, the senior officer first. As I expect, he steps forward and bows slightly. His grip is steady without smashing my hand. He winks and smiles at me. The second man, Petros, is shorter and seems quieter and less assertive. He looks away from my eyes—out of shyness, I think. My father has told me about this conversation. We have been expecting an invitation to join Alexander's campaign to the east in pursuit of Darius III. Two years ago, when Alexander first heard of the superior weapons the Persians had made for them by a Greek Jew in Memphis, he told his officers that he must get his hands on them. Through black market channels and battlefield scavenging, our swords, scythes, armor, and armaments made their way to the Macedonian troops. It is how Alexander gained Tyre, Gaza, and entry to Egypt. It is how his armies sent Darius III packing, much to our joy, and it is how Captain Kallias and Lieutenant Petros stand before us today.

Captain Kallias turns back to my father. "I can assure you that Alexander will make these things right for you, Barak. At the end of the campaign, you will be returned to your home, your family, and your business here. They will have prospered within the protections of the army we will leave behind, and the contracts for tools, equipment, and services that such an army always needs.

"Alexander has also suggested that you and some of your family join him on his trip to the seacoast, and then to the oasis at Siwah in the western desert. He wants to pay his respects to Ammon's oracle there. Can I assure him of your acceptance of his offer?"

My father holds up an indulgent hand. "I have no wish to keep you in suspense. Of course, I will accept the invitation to travel with your king. I would like to bring my oldest son, Nikola, and my daughter Aello – here with us now – along, if that meets Alexander's pleasure."

"We serve at the pleasure of our king," Kallias says, "but we will advise him of your interests in the trip, and that you have not rejected his offer to merge some part of your metalworking shop with our campaign."

"I think you can be more positive about my interest in your campaign than you indicate by your tone of voice," Father says. "I am very honored by Alexander's offer, and I can see much of benefit for myself and my family.

Thank you. I'm sure I will have an answer for you and for your king by the time we return from Siwah. Will I hear from you soon about trip details?"

"Yes. One more question for your daughter, if you don't mind. Alexander will want to know." The captain turns to me. "Do you have a particular interest in coming with us, Aello? Do you like to travel? Do you have an interest in Persian culture? Where we are going, we will confront our enemy's culture at every turn. An interest and a willingness to learn their ways can be of great value to us." I turned to my father for help in offering an answer. I am used to the hard work of the smithy, not to offering political niceties to superior officers of a foreign army.

"My daughter is very bright," Father says kindly, "and very curious. In addition to her studies she has been working with me in my shop since she was seven years old. She is very skilled in leatherwork and the making of shields. She has the design skills for flags and pennants. She seems to have a particular interest in understanding the strengths and weaknesses in the metals used in sword making, and in the practice of making tools and weapons out of iron.

"I cannot imagine going on such a war campaign as you describe, with such important responsibilities to the troops in your army, without having access to Aello's extensive practical knowledge and her skills at organizing our work."

"Oh, my dear sir. I am delighted with your answer," Kallias responds, "and I am sure my king will be as well. Our concern will be with her comfort and protection, I can assure you. Are you sure, Aello, that you will be comfortable enough in field conditions over a period of several months... even years?"

It is hard not to feel a growing passion in light of his question. Between our annual sandstorms, freezing desert nights, and the heat of the forge, I am no stranger to hard living conditions. I fold my arms over my breasts and say, "I am honored to come with you to Siwah, sir. What's more, if my job will be to help you destroy the

Persian army, I will sleep on rocks without blankets in winter. I do not enjoy my own words used in this way, but the Persians have driven me to them by their rapacity and the harm they have rained down on my friends, my family, and myself."

My father grins at me before he speaks again. "I will discuss all that we have talked about today with my family. You will all have an opportunity to speak with Aello at length on the trip to Siwah. You can judge for yourself about her maturity and ability to be effective in the field. Until I hear from you, then?" he asks.

"Yes. Of course. We will talk to you again as soon as we have the details worked out. Thank you for hearing us out."

***

I have worked in my father's metal and leather shop since I was a child.

I like to think I am pretty in the face, but I have grown muscular through my work. I like working with Tabu and the others in crafting the shop's products— even in the high heat of summer, when the furnaces cause the air to feel as though it were on fire.

My friend Tabu is Nubian. He is a giant of a man who was taken as a slave and emasculated by the occupying Persians as a youth. He is skilled in pounding heated ore into bronze and steel. Over the years, he has grown very fond of me and has become my protector.

He protects me from the other laborers who sometimes look toward me in ways that Tabu does not like.

Once, a Persian soldier sent to deliver an order for products to my father found me working alone in one of the storerooms. He was attempting to force himself on me when Tabu walked in. Tabu threw the soldier at the wall with such force that his head split open and he was immediately killed. Tabu admonished me to speak to no one about the incident.

Afterward, he told anyone who was curious that the soldier had tried to steal some of the valuable materials from a high shelf, then had stumbled and fallen to the stone floor. The Persians were so corrupt in those days that my father was able to persuade the soldiers sent to investigate that Tabu was a humble and simple man who could not lie. The Persians were skeptical, but their attitudes shortly improved when Father offered them each a sizeable payment to forget the whole matter.

I have never forgotten my friend's kindness and protection, nor that he likened me to the valuable, strong metals we keep on our highest shelves.

In many ways my life is one of repayment to this black giant of a man.

***

After our first meeting, Alexander again sends Kallias and Petros to inspect our forge.

The men stand in a corner of our crowded workshop in a large building on the Nile docks. Many workmen, sweating heavily, are busy hammering glowing pieces of metal. The heat is unbearable. The men are naked under the leather aprons we all wear to keep the sparks and hot flecks of metal off our bare skin as the hammers hit home. I am, too.

Father stands aside as my older brother, Nikola, hammers a glowing piece of bronze into a knife. Kallias and Petros look on.

"I believe you are generally familiar with the metalworking process," Father says, "but I want to make sure you have no illusions about the difficulties of creating a portable metalworking system. Such hot fires require several mature trees dried and rendered into charcoal. Charcoal is constantly fed into the fire, and the copper and tin ores that make up bronze are thrown in together in a clay vessel buried in the burning charcoal.

"If it becomes necessary to move, it will take us at least a day to have our wagons loaded and underway. It will take at least another day to set up again and fire our kilns at a new location. Add in whatever time it will take to move from the old site to the new. I believe the takedown and setup time will fall as we gain experience with the frequency of the demands for new weapons and armor, or the repair of the old ones."

Kallias nods. "I will carry that exact message back to Alexander. Now, I am going to have to leave this room. The heat is growing unbearable, and I feel the need to jump into a cold bath, lest I faint away."

Father grins knowingly. "We can accommodate you, if you wish. There is an outside pool filled with cool water near where my daughter and I keep our ledgers and do our design work. You and Petros are welcome to use it. Come with me." He turns to my corner of the shop. "Aello, come join us."

As I doff my leather apron and put on a light shoulder wrap to accompany my waistcloth, I hear him remind Captain Kallias that having me along with him is not negotiable. "She is not only one of my best craftsmen, she is also my right hand in the operation my business, and in our experiments with the making of steel."

"Please excuse my immodesty, gentlemen," I say as I join them. "It is the heat today."

Kallias seems unmoved as he smiles and offers greeting, but Petros seems embarrassed and distraught. He quickly looks away. "I am very sorry, Miss, " he says.

This amuses me greatly, and I smile in an attempt to put Petros at ease. I rise to shake hands with both men.

"We will use the pool," Kallias says. "Thank you for your hospitality, Barak. Before we return to our garrison, we would like to discuss two other matters of interest to our king."

"Of course," Father says. "Iva, my wife, may have some beer and bread to share. You are welcome to join us."

"Thank you. Petros and I will be very pleased to join you."

***

The two men rejoin us soon after, looking very much refreshed, and fall upon the tray of dates, olives, bread, and beer my mother has prepa "Our king asks me to inquire as to your skills with making steel," Kallias says to me as he inspects an olive. "Our experience with bronze weapons and armor is extensive, and our troops have learned many important lessons in their manufacture and use. We know that steel is lighter and will hold its edge much longer than bronze. Do you know of these things?"

"Another of my father's workers called Tabu and I have begun to explore the manufacture and use of steel," I say. "We have found much to recommend its use in the field over weapons made of bronze. However, it is somewhat more demanding in its manufacture. The iron ingots must be heated with coal, then hammered, cooled, then reheated, hammered, and cooled again. This process must be repeated several times to get rid of its impurities and refine its shape into a useable weapon.

"The process of shaping iron ingots into useful steel is physically demanding. We are trying to perfect ways to make high-quality steel in great quantities without killing ourselves with overwork. One of these methods involves quenching the steel in a saltwater bath rather than hammering, but we need more time and more experimentation."

I sit forward. "If you would like I can introduce you to Tabu. Once you see him and some of his work, I am sure you will understand something of the physical strength I think is required to make steel of weapons grade most efficiently."

"There is no need to meet Tabu now," the captain says. "If he is as good at steelmaking as you say, I'm sure you will want him to go with you on our campaign."

"Yes. There is another factor," I add, "perhaps most important of all. Iron ore is usually plentiful, but not always easy to find, or to remove from the earth, or to move from the site of a mine to the site of a smelting furnace. I have no knowledge of Alexander's proposed route, nor do I have any idea of the kinds or qualities of ores we might find.

"What I do know is that while we are advancing, your king must constantly judge our need for raw copper, tin, and iron ore with which to meet his demands for weapons. I am certain he will maintain reconnaissance looking for the enemy, and for safe routes through difficult areas. This reconnaissance must also include a rigorous search for the ore needed to make weapons." I punch a firm fist into the table to make my point. "This has to be an essential factor in all of his battle planning and campaign logistics. If he already has such a unit that he has confidence in, then I need say no more about it. If not, please advise me of what he thinks he may need, and I will prepare a proposal."

"My daughter speaks important truths," my father says in support of all that I have presented. Kallias nods to my father. "I will pass your concerns to Alexander. For now, perhaps I can suggest that your man craft a steel sword suitable for inspection by a warrior king. Bring Tabu and the sword along on our trip to the Mediterranean coast, and to Ammon's oracle at Siwah."

"It will be my pleasure to so instruct Tabu," Barak replies.

"We will depart to the north by ship within the week," Kallias says.

"Alexander has a great interest in building Greek shipping interests up to the point where we can displace the Phoenicians. For that he will need a major seaport on the Mediterranean Sea, and a strong navy to secure the commerce.

"We will spend time along the coast looking for a site to develop this port." He smiles fondly. "In addition to his many talents for wartime strategy and tactics, Alexander also has many skills in city planning and architecture. We will have an opportunity to see all of these played out as we work with him and think about a city and seaport of a size sufficient to handle the maritime commerce that we foresee."

I look at our guests and smile at the possibility of having so many things to learn from their king. Petros' look lingers for a moment. Then he smiles at me and looks back toward Kallias.

"You and your party should consider yourselves to be very fortunate," Kallias asserts. "Once Alexander is satisfied with his planning work for the new city of Alexandria, most of the army will return to Memphis.

Some will remain to provide security for our planners and surveyors at the site of our King's proposed new City of Alexandria. Our King will move by overland caravan to the west accompanied only by a small group of us. We think the trip to Siwah might take as long as ten days. We will be a fairly small group, though Alexander will have some of his best men, discretely armed, ahead, behind, and among us.

"The journey to Siwah and back means we could be away from Memphis for about two months. Can you make your preparations for joining us, and begin to arrange your affairs in Memphis for a possibly longer absence of two or three years? When my king asks for your statement on this, he will be very alert to any ambiguity or your lack of full confidence in what you tell us."

"Yes. We can be prepared enough to sound very confident of ourselves," Father replies.

"Good. My King, Petros, and I look forward to many conversations with you and your workmen..."

"...and my daughter," Father asserts. I bite my tongue. If it was not clear from the way that Kallias entertained my worries about forging steel on the warpath, then I don't know when it will be.

"Of course," Kallias says. "My King has already expressed his willingness to include whomever you think must go with us, and certainly the King understands that your daughter is very important to your work. Actually, it was he who took particular interest in her knowledge of making steel. She will be welcome. We know how important it is that we have a good working relationship with you and all your craftsmen."

He bends closer to us and speaks just above a whisper. "Our campaign will be hot, difficult, and dangerous. We have enjoyed great success in our battles so far, but I feel the success of our future campaigns may well depend on our adaptability to our surroundings as we get farther from our home territories. Our success will depend absolutely on the level of commitment to our cause by you, by each of your craftsmen, by your daughter, and by all who are both loyal to my King and technically skilled in the strategy, tactics, war-making details, and day-to-day minutiae in support of conquest.

"All will be challenged at our weakest points by our very capable enemies with their large and very well-equipped armies." Kallias pauses. "My King also believes this. You may trust my word."

Part 2

In January, we depart Memphis, down the westernmost tributary of the Nile, on two quinqueremes, the largest and fastest ships of Alexander's navy. Because of my country's warm welcome, Alexander feels no need to beef up security on the way to the small port of Rhakotis on the coast of the Mediterranean Sea.

The existing port is adequate for the level of commerce and shipping now moving through the eastern Mediterranean, but Alexander has shared with us, through Kallias and Petros, that he has bigger plans. A talented city planner and architect in his own right, he brings with him a delegation of planners, architects, and engineers familiar with building large structures on soft ground.

Tabu has come along with Father, Nikola, and me. He has completed his work on the steel sword that Kallias suggested back in Memphis. The sword is to be presented to Alexander for inspection—a ceremony also suggested by Kallias.

Tabu's presentation sword is simple. We agreed, in the forge, that Alexander is sophisticated enough in his choice of weapon that he does not need to be distracted by decoration or by extra, unnecessary weight. The steel sword is straight, and the blade is narrow, unlike the curved blade of the kopis Alexander once favored, which widens toward the tip. Tabu fought in the tribal wars of his homeland at the headwaters of the Nile. He has experience with weaponry, and with the death and disfigurement that a warrior could bring with a good weapon, but he intimated to me that he had never held a weapon that he felt to be such a part and extension of his body, that has such an instinctive flow as he swings it toward an imagined foe, as this sword. He is very proud of his work. We all are.

I have always wanted to learn all I can about making steel. As such, I helped Tabu with his work in any way that I could. At the end of the process, he asked my advice on some kind of message to be engraved onto the blade, a message that will be pleasing to Alexander. I was delighted that Tabu placed such confidence in me.

As we waited to embark from Memphis, I kept company with the shy Lieutenant Petros, at least on those days when he was not otherwise occupied with the business of Alexander. Together, we conferred with one of my most trusted and talented scribes about languages, phrases, and writing styles that would be appropriate for the newly named Pharaoh of Egypt. The scribe suggested a phrase and a style for a script to be written in Greek, along with a Greek engraver who will do the work on the finished steel.

Tabu was very pleased with the results of my recruitment work, and we scheduled the engraving to be completed before our ships set sail.

The trip to Rhakotis on Alexander's flag ship is short and uneventful.

The trip takes less than a day, and I spend those hours with Petros. I am growing increasingly fond of him. Even though we have spent time together and have allowed ourselves some intimacy, I understand his shyness and reticence. I do all I can to put him at ease.

***

At Kallias' urging, Alexander allows me to accompany his planning group on their two-day trek around Lake Mareotis, the natural harbor upon which the Pharaoh plans to build his city. Kallias tells me that I am to listen more than I speak, a demand that I am nervous I may not be able to honor.

"My King," Kallias says, "may I introduce to you, Aello, daughter of Barak?"

I bow low. "It is an honor, sir." The Pharaoh Alexander is younger than I imagined, but his youth belies a formidable confidence that I find enthralling and intimidating; even a little tactile. Beyond being introduced to each other, we do not get to speak, but I know that this face-to-face meeting is the greatest opportunity of my life. I decide to spend as much time as I can trying to comprehend all the things that are being discussed among these skilled and dedicated men.

There are so many new ideas that the architects and planners share, and I have so many questions, but I worry that interrupting their deliberations could ruin the welcome that Kallias has prepared for me.

It seems that Alexander envisions several permanent structures for his new city, including government buildings, a temple for worship of Greek and Egyptian gods, and even a giant library. "And naturally," I overhear a planner state, "the roads will connect the island of Pharos to the docks, starting with the shops."

I wince. It is my experience, working as we do near the docks of the Nile, that visitors to Memphis go directly to the buildings that are connected to them. Do they really want visitors to this grand city to visit its markets and government buildings before anything else? I am suddenly overcome with an intense desire to ask one of the planners about this.

Alexander's architect, Dinocrates, is huddled a few steps away with one of the other architects over a sketch of the city proper. I blurt out a little louder than I should have: "But the road connecting to the breakwater should go straight to the library and then to the government offices! It is a mistake to make the road connect the docks first with the shops."

The planner is suddenly embarrassed. He turns his eyes to the west. I see Kallias look at me with those piercing eyes raised to the sky.

Dinocrates turns to find the source of this disturbance. Thankfully, he recognizes me as a guest of Alexander and motions for me to come to him. He points to the layout. "Show us here on our sketch where the road should go."

I approach and bend to look. Once again I speak with more intensity that I intend: "Alexander has said that the library to be erected here will be the greatest in the known world. In their passion for knowledge people will come from all over the empire to see the vast arrays of papyrus and parchment, and the many scrolls to be stored here. He speaks of his dreams of the great poet Homer and all of his works in particular." I trace a new path on the sketch. "The road must go here. The people coming here by ship will not want to be taken to shops or government buildings.

They will want to be taken to the place of learning first—before anything else!" "But the road must also serve the interests of shippers of grains from the Nile, and the shippers of many goods from all around the Mediterranean Sea who wish to trade with the people of Egypt," Dinocrates replies patiently.

"The grains will not care how long it takes to arrive at their warehouses, but the seekers of knowledge will care very much if they are delayed in their quest, " I answer. "Build a separate road for the grain shippers here."

I trace another path on the sketch.

Dinocrates examines my two suggestions and turns to his counterpart to confer. "Your suggestions are very wise, girl. I will take them to Alexander and urge him to give consideration to your proposed roadways."

I rock back on my heels as he departs. Kallias catches my eye and gives me a gesture of approval. I'm thrilled by the fact that Dinocrates has listened to me. I want to tell my father, Nikola, and Tabu, of course, of all I have seen on this wonderful day.

***

"How was your time with the great man in the new City of Alexandria, my daughter?" Father asks as, breathless, I rejoin him two days later.

"He is a beautiful man, Father. He is not a tall man. Tabu would tower over him, but he walks as though he contains all the fires of your many forges, all throwing sparks with the sounds of all the hammers all at once," I reply.

"That is high praise. Will you be able to work effectively with him? It sounds like your first view of him has caused you to swoon," Father teases me.

"No, of course not. I think I know exactly what I am feeling—respect, though I admit I am less sure of myself in working closely with him. I know that anyone who gets close to Alexander can easily be consumed by him. That person could end up with no more value to herself or to others than the slag hammered off a piece of hot iron ore."

I sit by Father's side. "On the other hand, the intensity and heat of his personality are not all there is to the man himself. He is an intelligent and capable man of great vision. If I learned nothing else during the time we spent walking around the hot sands of Lake Mareotis, it is the fullness of his vision for his new city by the sea. On seeing the secure harbor and the safety of the approach from the sea through natural breakwaters, from that moment forward he literally carried an image of the docks and the whole city in his head. When he paused to consult this mental image, he only did so to instruct the architects and surveyors where each street centerline and foundation corner must be placed.

"I can assure you, though, that I did worry about my ability to think clearly when I stood close enough to hear his voice as he directed someone to move a pointer a little forward or a little backward. One of the architects was standing on the rocks overlooking the waves splashing onto the shore below while writing notes in his book. I'm sure it was an accident, but he suddenly dropped the book into the water several body lengths below, too far down to retrieve." Father's eyes widen, and I nod.

"Yes, the architect feared the worst. He sat down and put his head in his arms.

"I was nearby and saw a look of pure fury fly cross Alexander's eyes. I shook with fear for the note keeper. Even so, Alexander walked over to the man and put his arm around his shoulders to calm him down and get him back to work with a new book. After a few minutes, both stood up, the architect began making notes once again, and Alexander walked back toward us. As he passed by me, he winked then sent one of his warriors down to retrieve the planner's dropped book."

Father seems to consider this more deeply than I anticipated. "Perhaps he calmed down only because he knew you were watching him."

I shrug. "I am sure that is a possibility. He is the leader of one of the largest armies on the earth, and there are many stories of the terrible violence, beheadings, and disembowelments he has done to the opposing armies, even to his own soldiers who have become disloyal. Those stories do not reflect the character of a gentle or kindly man. I'm sure he will not be so forgiving with those he engages in battle, even those on his own staff if he knows he must make an example of them in order to avoid very ugly consequences."

"I am glad you told me these things, my daughter," Father says. "They make me all the more certain that we should join the Pharaoh on his visit across the desert to the oracle of Ammon. It will be important to our final decision on who will go east with his campaign. For now, let us prepare for the presentation of Tabu's sword. After that we must finish preparing for the trip to Siwah."

I step forward to hug my father. He is a good man, and I am glad to return to his side. "Thank you for taking the time to listen to me and for bringing me along on this journey. I wish I could capture your words as you speak them. I would paste them to my body so I would always have them where I can always see them." On the night before our departure, Kallias finally manages to get everyone together for the presentation of the sword. It will be a fairly small, fairly quiet gathering among some trees in a shadowed copse next to a stream. The presentation of a new sword hammered for a Pharaoh requires dignified simplicity rather than excess.

Petros has told me there will be no speeches, no formal introductions.

The Pharaoh walks up to a small stage under a tent and takes his seat.

He is dressed casually, though he does wear a bronze cuirass and his belt holds a short sword. His sun-darkened skin contrasts sharply with his curly, sun-blond hair. He is a handsome man, very sure of himself, though he sits as though he may need to jump into action at the slightest provocation.

He is very young, as I observed on the cliffs of Lake Mareotis—only twenty-four years old. Petros and Kallias walk behind the Pharaoh, and to his right. A four-man detail of soldiers in full armor and weaponry spread themselves along the rear of the small audience of favored officers and senior staff.

We stand in front of the short platform that Alexander has now ascended.

Barak, Nikola, and I are in clean white robes, with jewelry and headdresses suitable for an audience with the Pharaoh. Tabu stands at the center of the gathering, his forearms extended in front, his two huge hands cradling the sword waist high, a piece of white cloth draped over it.

Kallias whispers something in Tabu's ear, and Tabu, clearly overwhelmed with the dignity of the ceremony and his place in it, steps forward, approaches the Pharaoh, and kneels with his head bowed down, the sword extended toward the Pharaoh in presentation.

Alexander rises from his seat. "Please stand, Tabu. I am told you are a maker of fine steel swords. To a soldier like me, such a man deserves to stand eye to eye, even though your height puts your eyes a ways above mine. What do you have for me?"

Tabu seems afraid to open his mouth in response. He cradles the sword with his left arm and takes the cloth away from the sheathed sword with his right. He drops the cloth, but keeps his head bowed while he extends the sword to the Pharaoh. Alexander clutches the hilt with his right hand and pulls the sword from its sheath. Tabu drops his arms to his side. His eyes remain downcast as he backs away.

Alexander puts the sheath aside and turns to his left away from Kallias and Petros, swinging and twirling the sword with his wrist as he steps off the platform and moves away from the gathering. Suddenly, he swings the sword hard, first left, then right. He walks to a tree as big around as his forearm. He swings the edge of the sword, hard, at the trunk. The tree top falls. Before anybody can react, he continues in the same motion by swinging at the tree again backhanded. Another piece of the trunk falls.

He tosses the sword in the air and catches it with his left hand. In one sweeping motion he aims at another tree. The top of the other tree falls.

Again, he swings in a single backhanded motion, and another piece of the second trunk falls.

He turns and calls out to Kallias, standing among the troops there.

"Come face me. Be ready to fight."

Kallias does as he is bade by his King. He is an excellent swordsman, and I am told that Alexander often uses him as a sparring partner in swordplay. I imagine that these kinds of intense matches are how they keep themselves in fighting trim. They are also important in testing personal swords. The bronze, wide-bladed kopis held by Kallias is a personal killing sword, a standard design distributed among some of the infantry and all of the officers of the Greek army. By contrast there is this new sword hammered, personally, by Tabu into sharpened steel.

Alexander walks back toward the group, again lightly swinging and twirling the blued steel while he examines the length of it. He reads the engraving aloud: "Plato is dear to me, but dearer still is truth." He gives a smart grin. As he returns the sword to its sheath, he says, "A fine statement by my teaching master, Aristotle; a fine statement for a fine weapon.

Thank you, Tabu. I hope you and your master will agree to join us on our campaign to the east."

As Alexander turns back to face the approaching Kallias, now armed and armored, Tabu speaks. "Thank you, my Pharaoh. It is my great honor to serve you."

Alexander's first thrust to the chest is easily parried by Kallias, who then moves to block Alexander's next killing thrust low toward his belly. My breath is drained away. I sense my open mouth and think to myself that I must close it.

These two men are closely matched and well aware of the many practiced and usually successful thrusts and parries that they have had to both execute and also protect against. After a few minutes of intense swordplay, Alexander suddenly strikes, hard, at Kallias' kopis. The bronze blade breaks under the power of Alexander's stroke. The broken blade falls to the ground and lays there.

Alexander's theatrical nature breaks through as he thrusts his sword against Kallias' now undefended neck. He stops within a finger's width of raw flesh, and then lays the flat of the blade against the skin of Kallias' neck.

Alexander bends to pick up the broken blade of the kopis. I can see that there are several dents in the cutting edges, and, ultimately, a break where the cutting edge of the steel blade connected dents on the opposite sides of the bronze blade. I share an excited look with Tabu. Our weapon is clearly superior.

"An excellent test, my brother and friend," Alexander says. "This tells me all I need to know about this steel blade that our new friend Tabu has crafted for us." After thanking Tabu once more, Alexander departs quickly with his four bodyguards and Kallias.

Tabu seems embarrassed, but very pleased with Kallias' attention. He joins my family, and we walk back to our quarters feeling great pride in the fine piece of work Tabu has made.

The next day we arise early for the ten-day trip along the coast, then south to Siwah. The trip continues routinely along the seacoast, but the routine is not to last as we turn south away from the coast, and across the hot, dry desert.

While enroute Petros tells me that Alexander had given Tabu's sword to Kallias for safe keeping with his other prize possessions after the presentation.

"When Kallias came into the tent where we were billeted, he let me hold the sword. He then stood aside and urged me to swing the sword. I did so and nearly slash the wall of our tent. Tabu has burnished the sword to a dull bluish glow that imbues the sword with a powerful life of its own.

The engraving seems to give the living sword a voice.

Petros bends his head to mine and whispers into my ear. "I have never seen or felt or held a sword that comes to life in your hands and then speaks with such authority. Alexander will be very happy with it. He will be proud to have received it from your family.

Part 3.

Desert caravans are a wonder to behold. Though small with only 30 camels, Alexander's caravan on this first day is no exception to this wonderment.

Our desert guides are nomadic Berber tribesmen. They wear long, colorful robes wrapped around their bodies with multi-colored cloths around their turbans to keep the blowing sand out of their faces.

The Berbers can move across the desert without guideposts ahead and without leaving tracks behind. Alexander joins the Berber men walking more often than he rides. Not to be outdone, the men with Alexander, including my father and brother, also walk. I also walk. By their hand gestures and the words "Biya! Biya!" the Berber women ask me to come ride with them, but I wave their invitations aside.

Those who ride are mostly the Berber women, their small children, and some very old men. Pack animals, mostly burros, loaded with gear, provisions, and weapons, sometimes with children, make up the bulk of the caravan. The camels, burros, and other livestock have bells and colorful pom-poms and woven, multicolored leads attached to their halters and pack saddles. The Berber men often break into their native songs, and Alexander, enjoying the spectacle, commands his men to join in with Greek drinking and marching songs.

Four days south of the coast, a sandstorm forms in the west and moves toward us. It shifts, in the distance, in high, roiling brown clouds. The Berber guides tell us to stop and secure the material in the caravan against the wind and blowing sand. They will put up tents, heavily anchored by long pegs hammered deep in the sand, for the people. The animals will all need to be hobbled. Petros tells me that the guides offered Alexander a separate tent. He declined.

"He said he'd stay in the tent with his soldiers and our guests," Petros says proudly.

The wind and windborne sand blow strongly for more than a day. We can hear the sand blowing over the tents making a sound like an abrasive cloth. I fear the tent material will go weak from the abrasion and tear, but fortunately, they do not.

Eventually the storm abates. As we begin to crawl out from our tents, the guides confer with Alexander through Petros. Petros later tells me, as we continue our march, that they told Alexander that the landscape had entirely changed and that they may have lost some of the landmarks they rely on. Without landmarks they cannot guarantee that they can find the oasis at Siwah, and yet, we march on.

I watched as Alexander conferred with one of the Berber guides. I was exiting my tent, from the howling vacuum of fabric into a world of sand drifts. The Berber man's eyes popped at what Alexander was saying, and it was clear that the Pharaoh's youth and his belief in his own divinity gave him courage. I only pray, as we march on, that it is courage and not hubris. Petros says that we have food enough for another ten days, but that the supplies of water are poor. He guesses we have another five days under careful rationing.

Before we had our caravan packed to continue our travel south the Berber guide with Alexander had pointed in a direction, one that we are now following. Alexander nodded to him and then went to sit by himself for a time. Whether he was collecting himself or gathering the courage to lead in the face of a near certainty of failure, I'm not sure.

I took note of a flock of about a dozen ravens flying in circles far above the dunes ahead of him. They were pointing in the direction that the guide had indicated. I prayed that we might interpret this as an omen of good luck.

***

Four days later, just before dawn begins to lighten the sky, the first watering hole remains undiscovered. I wrap myself up in one of the Berber women's colorful wool robes and join Alexander outside the tents as he, groggy and thirsty, seeks the rising sun. I can tell that he hears more than sees the ravens circling and gabbling nearby, just as they have since we began our journey. I pay them more heed; I think they may be our salvation.

"Good morning, my Pharaoh," I say. "I have been watching the ravens.

One of them flies outside the main group. It looks at me from time to time." I pause, gauging the young King's reaction. "I think I can communicate with him."

Petros tells me that Alexander is not ignorant of such possibilities as ornithomancy. His mother, Olympias, told him he was destined for greatness, even divinity. Now that we have made him our pharaoh, he seems to think of himself as godlike. Indeed, his near-godliness is the very reason he wants to speak with the Oracle of Ammon in Siwah, just as Zeus had done so long ago.

With all of this in mind, I'm grateful that he says, "If the raven speaks you must listen and learn. Come back to me when you are sure you know his message."

"Thank you, my Pharaoh." I rise and walk in the direction of the circling ravens. When I near them, I sit down at the top of a high dune with my head bowed low, my robe pulled close in the cool morning air. Shortly, I feel the presence of another being. The raven that has been flying outside the group is now on the ground a few paces away from me, looking at me, making soft vocalizations. After a pause, the raven turns away, extends his wings, and flies away. I see a single white feather on his right wing. As the bird ascends into the sky, he turns again to look at me, then rejoins the circling ravens.

The flock flies in one direction and then descends among the dunes, almost out of sight against the dark sky a few hundred paces away. I look at the sky above the descending ravens. At the southern horizon, I see four stars at the end points of what appears to be a small cross. A few more moments and the stars will fade in the gathering light of dawn. I hasten to Alexander's side. "We must go in the direction of those four stars. We will find water there."

Alexander rubs his patchy, youthful beard in thought. "Those four stars are sometimes called the Southern Cross by my naval officers. That is a good omen. Thank you, Aello. I will direct my men and the guides to load their packs immediately and move toward those stars. You must remain here and search the horizon for landmarks under them that we can follow when the stars fade away in the dawn. I will instruct two of my men to stay a distance away from us on either side, on top of the dunes to look out for signs of water. You must concentrate on the location of the ravens.

If they move in a new direction, you must find me and tell me immediately."

"Yes, my Pharaoh."

Later that day, the travelers discover water. The guides say that the storm has uncovered a new and plentiful watering hole on the route to Siwah. They are pleased to see fresh water still gathering in the bottom of a shallow depression.

Alexander orders a day for us to recover and regroup, and a few days after that, the caravan arrives at the oasis in Siwah. Alexander has his meeting with the Oracle. To no one's surprise he will not speak of his experience there. The only evidence of a change in his attitude following the meeting is the purchase of a translucent blue/violet stone about the size of a man's thumb. It is beautiful. At first, I thought the stone might be Lapis Lazuli, a stone found only in the eastern mountains of Afghanistan far beyond Mesopotamia. A stone much favored by the Egyptians. As I looked more closely, I saw that the blue/violet color was more translucent than the sky blue of most Lapis and there were no flecks of gold.

As he holds it up to show his closest advisors, his face transforms into the first broad smile that I have seen since we began our journey at the port of Rachiotis a few weeks ago.

Part 4

It has been nearly two years since our ordeal in the desert, since Alexander's visit to the Oracle at Siwah. My pharaoh never speaks of his conversation with the Oracle... not with anybody.

In the intervening years, he caught up with the fleeing Darius III, finally, in a tough battle at the Caspian Gates. Darius was already dead, whether killed by his own people or by his own hand, we are not sure. All I know is that the battle was a rout worthy of the immortality of scrolls and legends. My father's tent was behind Parmenion's left flank, and we endured a small skirmish of our own to keep Persian forces out of our smithy. Alexander honored Darius as a valiant warrior and buried him with full military honors.

Now victorious, the older Macedonians in Alexander's army wanted to go home. In their minds, the war was over. In

Alexander's mind, and in the minds of younger soldiers loyal to him, Darius' army was in total disarray.

The way was now clear to roll up the rest of the Persian empire and take the Greek army all the way to India. In that way the Greeks would have an empire even greater than the Persians. I often wonder if this is what the oracle foretold, or if it is how Alexander interpreted it. Either way, we were on the move to the east again.

We approached the great capital at Persepolis that winter. The Persian capital lay abandoned before us. Our occupation of it and the resulting inaction over five cold months of winter was a terrible time. Alexander had taken on some of the attributes of a too-proud conqueror. Worse were his affectations of Persian habits and demeanor. He occasionally wore some of the fashions worn by Darius III and his sycophants at court.

Though this time was difficult and troubling for Alexander, I will never regret it. It led to a more intimate relationship between me and Petros, and we quietly moved into the same tent near my father's forgery. The smithy kept us warm through that harsh winter.

When spring came, Alexander was ready to move on to India, but first he would have to secure Bactria and Gandhara—a place of high mountains and regional tribes led by strong warriors. In order to establish the influence of the Greek city-states, he would need to keep the Macedonians with him and willingly under arms with his leadership.

My father has remained closely attached to Alexander's army since we departed Memphis. He has found favor with Alexander and his closest military advisors. With daily forge operations under Nikola's supervision, Father has been free to work closely with Alexander's operations planning staff to find sources of ore and other materials necessary for the manufacture and maintenance of weapons, armor, and other battle gear.

Increasingly, Alexander seems to grow firm in his belief that he is the son of Zeus. This belief has been growing since the Egyptians named him pharaoh four years before and has intensified with the successful battles at Gaugamela and at the Caspian Gates.

As we followed Alexander's army closer to India, my bond with Petros deepened. I became pregnant and gave birth to our beautiful son, Balio.

Together, as a new and loving family, Petros and I think about our future.

We are not certain that it lies with this army for much longer. As the ethnicity and language of the local armies drawn under Alexander's banner increases, though, Petros' work as an interpreter has become increasingly important. In the negotiations needed to persuade conquered armies to join us, Petros has become almost inseparable from his King.

Whenever he goes with Alexander or his staff on a diplomatic mission, Petros asks, usually through his friend and colleague, Kallias, to have me accompany them. Petros knows that Alexander will want me to evaluate weaponry problems, issues, and needs. We believe that he agrees to this because he is showing favor to two of his personal favorites. We are grateful for this. We have never weakened in our public support for his decisions and his policies, and never fail to tell him privately what he needs to know—no matter how uncomfortable the realities behind them might be.

***

It has been a long day for Alexander. I am conscious of this as I approach him in one of his all-too-frequent meetings. He must confer regularly with his senior staff regarding the establishment of a military garrison that would be at the economic heart of the new city of Alexandria in Arachosia. In his vision of empire these garrisons will permanently support his armies through expected growth that is to include all of India and much of the territory north of the Hindu Kush mountains. He has had no problem finding volunteers to settle in these garrisons. Many of his most loyal soldiers, weary now of the years of war over harsh terrain, welcome his offer of land and a significant share of the spoils of war.

We are among these volunteers.

"Petros and I have decided that our lives now belong to our son, my King," I say as we explain our plans to him. "There is no place for him on the long, hard trail over the mountains that you foresee over the coming months. My father, who began this journey with you four years ago in Memphis, is now, himself, old and tired from the effort to keep your soldiers well-armed. Even so, with the help of me, Petros, and Tabu, we believe we can continue to supply your army with significant quantities of weapons from the shops we will build, with your support, in this new city.

"Alexandria is rich with resources. Abundant fresh water from the mountains and from the river that flows through here from the foothills will enable a great agriculture. When your soldiers who agree to remain here turn their hard-won skills to these resources, they will create a paradise on earth. I and my family believe that our knowledge and skills with metalworking and our understanding of your vision will be important in helping guide this rough land into a new seat of power "We seek your support in this, my King. We are available to work with you and your city planners and architects to help make this shared goal into reality just as you and I did two years ago to layout the first City of Alexandria on the bright shores of the Mediterranean Sea. May we have your thoughts on this?"

At first Alexander seems taken aback by our bold proposal. I hope that it is clear that I, who have always been a relief from the intense and often destructive wrangling of his lieutenants over irrelevant details, have thought deeply of this and have discussed it fully with all who will be affected.

"Leave me now," Alexander says. "You have given me much to think about, and I am sure we will have more to discuss as I prepare for our expedition north. Our time for travel in the high mountains grows short as winter snows approach."

***

As the military commander of an army in the process of conquering much of the known world and, now, extending the limits of his empire to encompass all of India, Alexander knows that such considerations cannot be given much weight. As his mother always trained him, and as he has always done, our King and our friend seems to know the decisions must be made and the march north commenced without pausing to consider possible alternatives.

As Olympias always told him—as he is the son of Zeus—all errors of judgment will eventually be forgiven to a god.

END

Faith

Spring, 20 CE. In Jerusalem, the impetuous Gila and her older sister, Ziva, meet Jesus of Nazareth on a quiet street corner near the lower market. Later, under clandestine circumstances, the girls and their family are taken into domestic slavery in the family home of a Roman centurion.

====

While Gila is still a child, her family is taken from the life they have known in Bethlehem and now in Jerusalem. They are taken into domestic slavery in the home of the Roman centurion Quintus Caecillius. Gila knows she must learn to curb her independent, impulsive, occasionally disruptive nature, though in light of her enforced servitude, she finds this to be a nearly impossible task. Her only hope is to allow herself to indulge her independent nature in ways that are not obvious to Quintus or to the other members of his household. She starts by reading every scroll she can find in his library and yearns for the day she can return to her homeland, hoping for another encounter with the exorcist and miracle worker, Jesus.

Ziva is more artistic and less effusive than her younger sister, Gila. She has learned how to put up with Gila in her most disruptive moments, but she also knows that her own patience is not unlimited. Next in line for Gila to annoy is Elias, a brother older by a few years than both girls. Elias likes his duties working with his father, Avraham, taking care of the stables for the horses that are being trained by Quintus for sale to the Roman army.

Part 1

Ziva suddenly calls out, “Gila! This way! Quickly!”

The younger Gila runs to her sister’s side to see what she is yelling about. When she sees, her eyes light up. Around the corner, in one of the open shops in the bustling market of lower Jerusalem, a young, bearded man in a light gray robe is talking quietly to a crowd of people. The small audience has turned its ear to the soft words being spoken to them. Gila preacher from Galilee known as Jesus of Nazareth. Her father has worked as a craftsman in the Galilee and speaks often of the time he went to temple and Jesus was there. Now, she and her sister have found him, by luck, in the Passover marketplace.

The name Jesus is common. This Jesus is often known as the Nazarene to set him apart from the many prophets, exorcists, healers, and aspiring messiahs who work the crowds in Jerusalem during Passover. Because the preachers from Galilee are especially vigorous and imaginative in their presentations, shoppers will almost always stop and listen if for no other reason than to be entertained on a warm spring day.

This Jesus is well known because his good reputation for healing and exorcism has preceded him. Gila stands on her tiptoes and strains to hear his voice. The market is crowded, so it is very hard to hear his words unless she moves closer.

Jesus the Nazarene speaks of the one God in ways that cause his listeners to bend their bodies and turn their ears toward his voice. His voice is clear and modulated across a range of tones depending on the size of his audience, and on what he wants to emphasize to his hearer. He speaks to each person directly. He engages with them as individuals while they speak. He is alert to their questions and thoughtful about his answers.

His eyes and those of this listener connect in a way that impresses Gila with the power of the bond between the two, even if it is only a temporary one. In spite of her excitement, Ziva is worried. As Gila pulls on the hems of robes and steps on sandals to get closer to this man, Ziva knows she should not have brought her younger sister to this place full of excitement and strangers without their parents. Jerusalem is a Roman province now, and armed soldiers dressed in leather and metal patrol the streets with dour faces, looking for Jews to bother.

But today is special. It is spring, with all the greenery and flowers bursting through the dusty earth, and it is Passover, a time to celebrate their people's emancipation from slavery in Egypt a millennium ago. Ziva, Gila, and their family came from their home in Bethlehem to the main Jerusalem market in the lower city to celebrate with their friends from the nearby towns.

While their father Avraham chatted with his fellow leatherworkers in the market, the girls snuck off to amuse themselves. Ziva knows how to find the house of the relatives they are staying with, so she knows she's not giving her parents much concern. Avraham had alerted his girls that the Nazarene might be in Jerusalem during this Passover, even though the distance between Nazareth and Jerusalem is nearly sixty miles, a six- to ten-day journey by foot.

"He will be in Jerusalem this Passover, " their father had said, even as their mother, Orli, rolled her eyes. "I am sure of it."

Clearly the Nazarene's work as an exorcist, healer, and listener—but most particularly as a prophet who sees beyond the present day–has begun to resonate beyond the shores and villages of Galilee. Jesus' eventual visit to Jerusalem had become inevitable in the minds of those who know his work, and this Passover was the time for it.

Avraham describes the Nazarene as a fisherman, carpenter, and day laborer. Some part of his work, as with all young Jewish men in Galilee, involves rebuilding the Jewish regional capital of Sepphoris. As construction work at Sepphoris neared completion, construction of the new Roman capital on the western shore of the Sea of Galilee began to speed up.

Two years ago, the architect and developer of the Roman capital, Herod Antipas, named the site after his friend, mentor, and patron, the Roman emperor, Tiberius.

Avraham knew these things because of the many weeks and months he spent in Galilee around that time, earning good money as a leatherworker on the various construction sites. He had visited many small communities and had had many opportunities to hear Jesus speak. However, his obvious regard for the Nazarene teacher has caused strains among neighbors and acquaintances of their Jerusalem hosts who are otherwise inclined to be friendly, even to his two girls, even in their most turbulent moments.

Ziva shakes her head at her sister's antics. The latest hem she's pulled on belongs to one of the magistrates. He does not look pleased with Jesus' teachings or with her sister's mischief. Gila's curiosity often gets her into trouble with adults as she asks questions that might not otherwise be thought suitable for a child not yet ten years old. From her earliest age Gila would stop and stoop to turn over rocks in the streets of Bethlehem, to push away weeds so she could pick up bugs, and to dig around in loose soil to see what treasures lay beneath. When she would present nearby adults, even strangers, with a squirmy bug or slimy worm, some shrank away and insisted that she leave their presence or have demanded that her parents retrieve her and take her away. Ziva rolls her eyes—this is a job that usually falls to her.

Gila comes close enough that her sister can grab her arm. "Mind your manners," she whispers. Gila nods, though her eyes have a faraway look that suggests she won't be paying the command much attention. This day is one among many in a period of occasionally tense relations between the Jews and the Romans. Conflicts break out over the question of the one god of the Jews and the many gods of the Romans.

Tiberius, like Augustus before him, has no particular antipathy toward the Jews, but he has instructed his bureaucrats in the provinces to, once again, begin questioning Jews about their personal relationships to the many Roman gods. In response to these instructions, Roman soldiers have been confronting Jews on the streets who are known to participate in high holy days.

The soldiers pose questions that could provoke answers by individual Jews that might, or might not, be acceptable to their interrogators. Depending on the encounter, a soldier might ask about one of the minor gods. Since the Roman gods number in the hundreds, all the interrogatee can do is apologize for his or her lack of knowledge and then go on to promise to learn all that could be learned about whichever minor Roman deity had been the subject of the interrogation.

By this simple pledge, most Jews are released from further interrogation.

Some of the community leaders are concerned, though, that some have been taken into custody following these interrogations, and no family member or advocate is allowed to speak with them afterward. There are rumors in the market that some have been taken away to feed Rome's insatiable appetite for slaves to fuel its expanding empire. Even more important among the upper classes in the City of Rome itself are the constant needs for teachers for their children, and for slaves with skills in cooking, serving, entertainment, and household administration. In these domestic skills Jews are often most favored.

***

Jesus has now begun to walk toward the marketplace. When only a few listeners are surrounding Jesus, he often prefers to walk with them through the busy streets. Many of the shopkeepers and their waiting cus-tomers wave and express their greetings to the preacher from the Galilee. Jesus walks with his hand raised to acknowledge all the greetings, and often tips his head to those who are known to him.

Ziva and Gila are torn between following the group and leaving to find their parents. At that moment Jesus smiles in their direction and motions for them to come to him. Gila clasps her hand in her sister's and pulls her forward.

When they arrive at his side, Jesus crouches down to their level, and says something in a language that they don't understand. Ziva and Gila share a look, and Ziva shakes her head to indicate that she does not speak Aramaic.

"Do you speak Greek?" she asks. Jesus nods yes, but before he can say more, Gila announces, "Father says that Jesus is a good Jewish man, so he must know how to speak Hebrew."

Jesus suppresses a grin, then nods slowly at her. Gila can feel her face flush red. He then speaks in a rough Hebrew that he struggles with.

"Who... what are your names?"

"Ziva and Gila, daughters of Avraham the leatherworker and his wife, Orli."

"Ah," Jesus says, "I believe I know Avraham. He travels to the Galilee from time to time?" They nod. "He is a kind man, and wise." He looks Gila in the eye. "I have thought well of his words in synagogue."

He sat back on his heels and looked around the market. "Perhaps you should try to find your parents while it is still light. The evening meal approaches, and I'm sure your mother will be getting worried."

They both nod. "Peace be with you," he says by way of farewell. "May we meet again, next year in Jerusalem." He turns back toward his group of listeners and starts again to walk toward the market.

The girls stand briefly, awestruck. They say nothing because he has stolen their voices with his words of wisdom and his respect for their father. They look at each other again, and begin to run, slowly at first, and then as fast as they can, laughing and screaming through the narrow alleys and crowded spaces of Jerusalem's open air markets. As usual, their disruptions cause many bumped and ruffled adults to swear at them and condemn their parents.

Suddenly, as they turn to run down another alleyway, they see a group of Roman soldiers speaking with one of the shopkeepers a few dozen paces in front of them. Now, the soldiers and the shopkeeper are all looking at them.

One of the soldiers appears to be in charge of four legionnaires near him who are of lower rank. He is not an officer, but by the extra flash of red cloth on his shoulder, he is probably a decanus of the second cohort of the tenth legion. His name is Kanutus. The girls know because they have met him before. He steps away from the group and walks toward them. The ranking legionnaire speaks a rough dialect of Latin that suggests a poor upbringing and a lack of formal training in the language. Possibly he grew up on a farm near Rome, Ziva likes to think. Since Latin evolved from Greek, she is able to understand what the soldier wants to say, even though he does not say it very well in Latin. She and her sister turn from his threats and start to move away.

"Stop! " the decanus yells. "Are you the ones making all that noise?" They are terrified, and aware that the decanus is the one largely making the noise at this point, but both nod yes.

"Where are your parents?"

Ziva replies with the name of a nearby neighborhood known for the many Jewish families and storekeepers who live in the crowded apartments there.

"We will be in that neighborhood later today. What are your parents' names?" he asks.

Ziva's voice quavers. "Avraham... Orli." Her lip begins to quiver.

"Tell your parents that we will want to talk to them about your bad behavior. Now go along home, and don't run or scream like the mean and disrespectful children that you are," the decanus says.

They walk quietly past the other soldiers and past the shopkeeper who looks angrily in their direction. By making a scene and by giving up the names of their parents and their neighborhood, they have brought the attention of the Roman government down on their people and their family.

Neither of the girls talks again until they are in their own neighborhood, only a few steps from the shopkeeper's store and the apartment where their family is staying.

Orli turns from cutting carrots for dinner to greet her daughters. "It is getting dark. I am glad you are home. Where did you go, and what did you do today?"

"We saw Jesus!" Gila shouts, unable to contain herself. "He was in the market with some of his friends. He talked to us!"

"What did he say?" Orli wants to know. "Did he tell you why he is here?"

"He said he hopes to see us again during Passover!" Ziva chimes in, nib-bling on a the seeds of a pomegranate. "We must come back to Jerusalem next year! Oh, please, Mother, can we come back next year?"

"We will have to talk to your father about that," Orli says. "I don't know how this visit is going for your father. The Romans seem to be very curious about whether we like their gods, and which ones we like best. Your father tells them there are too many, and he gets confused. He always tries to convince them that they should think about the one God of the Jews." She pauses and looks out the window. "I worry when he does that."

"Why, Momma?" Gila asks.

A shadow crosses Orli's face, but then she shakes her head as though to clear it away. "Come help me chop up these eggplants, my sweet ones. We will talk later about all of this with your father. For now, I want to know that you know your Roman gods well enough for the street. Who is the god of the hearth?"

"Hestia!" Gila cheers. "That's an easy one."

"All right, then what does Neptune do?"

"He's the god of the sea. His brothers rule over the heavens and over Sheol below."

"The Underworld, darling," Orli corrects Ziva. "Romans don't say Sheol." She turns around, hands on her hips. "All right, here's a stumper: who's Adiona?"

The girls share a confused look. This certainly isn't a god from the major pantheon, but they know that their mother asks because the Romans are not likely to give them any easy answers. They have been schooled to answer that they do not know, if that is the truth, but to promise that they will learn the minor gods well. They are children, and it is hoped that this will give them a free pass an adult may not receive.

The answer comes to Gila like a lightning bolt. "Oh! I know! I know!" She waves her arm, nearly knocking over the bowl of chopped eggplant. "Adiona is the goddess of safe return. She ensures that children make their way home."

Orli grins and puts down her kitchen knife. She gathers her girls into her arms. "Precisely right."

Bathsheva, their Jerusalem hostess and friend, walks into the small kitchen. She is carrying a bottle of wine, as well as some olives, garlic bulbs, a tomato, and bread in her shopping bag. "Shabbot Shalom," she says. "Is dinner ready?"

"Shabbot Shalom. I'd like to wait for Avraham. He should be here soon. We can have a glass of wine in the meantime. Where is Aharon?" Orli asks.

"He is at synagogue with some friends. They have been arguing over the meaning of the same part of the Torah for years." Bathsheva speaks in the lowered voice of a man: "'Does Avraham have the right to question God?' I am sure I do not know. After all these years of argument they don't seem to know either." She looks at Ziva by the stove, who is starting a fire to cook the soup. "Since that Nazarene boy is here, maybe he can clear up their points at contention, though I don't know what they will find to argue about after that."

"I have listened to Avraham discuss the two times in Galilee when he and Jesus attended the same synagogue. I have not heard anything from Avraham that tells me that Jesus spends much time resolving points of contention," Orli says. "I understand that Jesus cannot read or write Hebrew, but Avraham says he doesn't seem to have any difficulty picking out elements of flawed logic from the inflated ramblings of the Sanhedrin. But..." She frowns, looks away, and shakes her head. "I'm sure I talk too much."

***

The newly promoted decanus responsible for patrolling the public market in the lower city of Jerusalem this day was told to stay outside the door while waiting to see his superior officer, Quintus, the centurion primus ordinis of the Second Cohort of the Tenth Legion. He has come, at Quintus' request, to report on the movements of the charismatic Nazarene around the streets.

Decanus Kanutus was handpicked by Quintus for duty in the Jewish capital of Jerusalem. He is a decorated legionnaire who fought and was wounded during the purge of the Jews from Rome in the previous year. He is loyal to Rome, to the Roman gods, and to Quintus himself. Quintus knows Kanutus can lead a squad of legionnaires into battle and would be effective as an enforcer of Roman law in the contentious Jewish capital. When Kanutus arrived in Jerusalem, Quintus promoted him to decanus.

Quintus also knows that Kanutus, once engaged in a face-to-face knife fight, will not stop until he has killed his opponent. During the purge, Kanutus watched, helpless, as his brother was killed in a terrible way

by one of the Jewish rebels. When Kanutus was able to break free of his own struggle, he followed the Jew, confronted him face to face, and with a single, quick, slashing move of his pugio, spilled the man's guts to the ground, then left him to die in the hot, dusty street. He therefore knows to keep an eye on Kanutus and his temper in this Jewish capital.

Quintus is on the balcony of his temporary offices on the Mount of Olives. He is facing the setting sun as it casts the city and the temple before him in growing shadow. He is speaking with Tribune Theodorus, who has come to inspect the work Quintus Caecillius is doing in Jerusalem. When Decanus Kanutus is finally motioned into the room, Quintus greets him with a smile. "Tribune Theodorus: please meet my decorated and very able Decanus Kanutus. He has been down on the streets these past many days and has much to report. Perhaps, if you will give him your time and attention, you can report back to Legate Kaius in Caesarea with much fascinating detail on our work here."

The tribune turns toward the decanus. Kanutus' knees are a little weak in the presence of these two legion officers. Because of their time together in Rome, he feels a powerful allegiance toward Quintus, but he knows nothing of the tribune, except, as is well known among the ranks of legionnaires in the field, a tribune is a political officer, not a fighting man. As such, he cannot be trusted under any circumstances. Since he will never appear on the field of battle, there is nothing that can change Kanutus' opinion of this tribune, nor of any other tribune.

"Hail to Caesar Tiberius." Kanutus snaps a formal Roman salute toward the Tribune.

"Very good, Decanus," the tribune says and returns the salute.

"Tell the Tribune what you have learned of the Nazarene," Quintus says. "Tell him the same things you have told me."

Kanutus nods and clears his throat. "I don't think the Nazarene is here to cause trouble for Rome. He talks to the people who know of him, or who have heard of him because of his reputation as an exorcist and healer in the Galilee. He stays away from large crowds; he stops talking and walks away when too many people gather round to hear him. I think he will eventually be a bigger problem for the Jews in the temple than for us."

"He is well known and well regarded throughout Galilee," the tribune says. "I have met him myself, and some of my staff have sought healing from him. The Legate is concerned about the effect he is having on the population in Galilee and is very curious to know why he has come to Jerusalem this Passover. Is this his first time here?"

"By everything I have found out about him, yes, it is his first time here," Kanutus says, "but, I think he intends to return."

"Why do you say that?" Quintus asks.

"Because he told two little girls that he met in the market that he hoped to see them at another Passover soon."

The tribune and Quintus share a look. Kanutus hopes they are not doubting his information merely because he overheard it from two chil-dren. "Are these little girls, or any Jewish civilians, inclined to confide in you in some casual way when you are in uniform, Decanus Kanutus?" asks the tribune.

"No, Tribune, but their parents, in hopes of avoiding trouble with Rome, are cooperative and eager to please when they do not see an immediate threat. I knew that when I approached them in their homes at dinnertime later that day. I was discrete and spoke to their father as one father to another, concerned that certain bad behaviors among his children should not continue."

"What reason did you have for going to their home?"

"Their two children, the two little girls, had been in the market and had met the Nazarene. In their excitement afterward they began running and screaming through the streets, disrupting the shops. I had a bit of business of my own in the market for some repairs to my uniform when they approached. I told them I wanted their parents' names, and they gave them to me. The oldest told me where they were staying.

"It turns out," he continues, "that their father, Avraham, an observant Jew from Bethlehem, has a reputation among the local legionnaires for not responding to questions about the Roman gods with enough seriousness. He is on a list of people Governor Gratus believes need to be watched."

"Governor Gratus causes a lot of problems for himself with the Jews when he dismisses the chosen leader of the Jewish Temple and replaces them with his personal lapdogs like Ishmael," Quintus says. "No wonder he is worried about Jews. I think Gratus may have partially solved the problem himself by accepting the appointment of Caiaphas. At least so far." He cracks his knuckles contemplatively. "This sounds like a political problem among the Jews that will take more time to either resolve itself or mature into something that Rome will need to be concerned about. But that time is not now, and you and I, Decanus, need to think about, and prepare for, our return to Rome.

"I will speak to my superiors in Rome, Tribune, about the problems Gratus has created for himself by his frequent removals and appointments of temple priests. As I said, it may have been partially corrected with his appointment of Caiaphas, but the actions of Gratus need close watching. I hope that you will inform the 10th Legion's legate of what you have learned here today. I hope you will express to him, also, my belief that there is no immediate cause for concern. I'm certain that my superiors in Rome will communicate any further information on this to Legate Kaius in Caesarea whenever it might become necessary." He spreads his arms to envelop the two other men. "For now, please make yourself at home in my home, Tribune. Decanus Kanutus and I have some operational business to discuss, then we can all sit down to dinner."

Quintus waits to continue speaking until after Tribune Theodorus has left the room. "You have done very good work on this, Decanus Kanutus, but I think your work has only just begun. I want you to somehow arrange for me to get a closer personal look at Avraham and his family. Since you are now acquainted with him on a personal level, a chance meeting in the market might serve my purposes very well."

"This is a surprise request, sir. If I may inquire, what are those purposes?

"You have told me that Avraham has spent many months plying his leather trade in Galilee and is probably better acquainted with Jesus than most other Jews here. I'm thinking it might be good to get a man like Avraham and his immediate family back to Rome to help us educate our-selves and our superiors about these Jews and their beliefs."

He pauses and grows thoughtful. "You don't need to discuss this with anybody else, but I believe that their belief in one god will corrode the foundations of our faith, and our beliefs in our many gods. Eventually the Jewish faith in one god will replace them. Rome needs to understand the political force these Jews represent, and must begin preparations for dealing with the consequences of their use of Jewish power."

"You may count on me, sir," Kanutus says. "Despite what we faced in Rome last year, I have to admit to a certain admiration for the Jewish faith, and the persistence of it among people who have little else of value that they can claim, though among my fellow legionnaires, sir, those beliefs and the threats they pose to their personal gods do not usually sit well."

***

The day that Quintus Caecillius and Avraham meet dawns warm and sunny. Quintus is in a corner of the open-air square just off the crowded market. He sits at a table near a large shade tree, sipping a small mug of wine. His uniform is casual, designed to show that he is a field commander in the legion, but with no obvious insignia of rank. Except for a short sword in his belt, he is unarmed.

He tears off a piece of bread and dips it into a shallow dish of chopped olives, oil, and spices. Quintus takes note of a group of ten or twelve ravens hopping around a piece of bread on the street a few steps away. One of the ravens, an odd-looking fellow with a white feather on top of his right wing, looks up to take note of Quintus, and hops sideways toward him. Once able to secure a piece of bread from the table, the raven hops off.

Decanus Kanutus appears at the edge of the square. With him is the subject of the previous evening's conversation: Avraham, the father of Gila and Ziva. Quintus motions them over to his table.

Decanus Kanutus approaches and offers a formal Roman salute. "Good morning, sir. May I present Avraham. Avraham was working with the leather smiths this morning. At my invitation, he has agreed to come with me to meet with you."

"Very good, Decanus." Quintus motions Avraham to a chair at his table. "Please sit down, Avraham. You may return to your duties, Decanus. I will find my own way back to the Mount of Olives."

"Very good, sir."

The other man looks wary, so Quintus says, "Please relax, Avraham. This is not an interrogation. Neither I nor any of my troops mean any harm to you or your family. I do understand that some of the legionnaires in the market are concerned about your casual attitudes toward our gods, but that is no part of the conversation I want to have with you now. I have always made it clear that I will not tolerate any complaints of harassment by any legionnaires over conflicting Jewish opinions about Roman gods.

"So. For purposes of conversation, do we have at least this temporary understanding?"

"I feel very disadvantaged here. Why have you asked me to speak with you?" Avraham asks.

Quintus pauses to gaze at him. "I don't see any point in wasting your time. I will be leaving Jerusalem within the next two months, and I am looking for a Jewish family familiar with the Judean and Syrian provinces to come to Rome with me and work as household staff to myself and my wife."

"Why don't you simply take us as slaves during the dark of night? That is the usual Roman way, is it not?" Avraham asks.

Quintus cannot be sure if he detects malice in the man's tone. He proceeds with caution. "Perhaps it will be necessary to take you to Rome as slaves, or so it will need to appear to others, but that is not my preference. I can assure you and your family that you will be decently compensated while you work for me, whether as slaves or free. If you work well, we can discuss the return of you and your family to Jerusalem after some period of service, and I can personally guarantee some measure of status and some amount of compensation to help you resettle here again, or in Rome if that becomes your wish."

"You are making a very generous offer," Avraham says. "I have no idea how to respond to you, and I have no idea why you are making such a generous offer to a total stranger who, by his culture and upbringing, may have hostile intent toward you and everything you hold dear, and who could readily lay in wait for an opportunity to kill you... and your family."

The two men pause and take in the other's countenance for a moment. Avraham, a practical joker at heart, is unable to keep a straight face for longer than that, and, as he looks away his frown cracks into a smile. Quintus, relieved that it appears he will not have to take offense as any good Roman should, also smiles. He looks into the eyes of Avraham as he composes himself and offers apology.

"I apologize to you sir. I have taken poor advantage of your gracious-ness and hospitality. Please forgive me."

"Accepted. Decanus Kanutus said you were a reasonable man," Quintus says, "but very sharp and very persistent. I will make a guarantee to you now that your return to Jerusalem with your family is assured, though we will need to negotiate a date... perhaps three or four years into the future. When you return, you and your family will enjoy considerable status among your peers, in whichever ways you and your peers define that term. You have my personal assurance of that.

"Let us leave aside the questions of motive, intent, and trust for now. Instead, let me describe the work I have in mind. I promise you, when the time is right, that I will tell you all that I have in mind as reasons for making this offer, and for making the offer to you in particular. First of all, I'm sure you realize that men in the upper reaches of Roman society often have extensive Jewish staff in their households, and that they are usually treated quite well, whether slave or free.

"My estate is just outside Rome on the way to the port of Ostia. I have been away from Rome and away from my estate for over a year. For now, my wife, Cornelia, is supervising others who are taking care of the properties, but when I return, they will also return to their own properties, and I will need a staff to work the property right away. If I don't bring staff back with me, I will need to recruit in Rome, and that will be very difficult and very time consuming at this time of year. Worse, I fear that I won't find enough good staff to turn the soil, plant my crops, and husband my stable of horses. This worries me a great deal.

"So. My first three reasons are relatively simple: property maintenance, spring planting, and animal husbandry. Decanus Kanutus tells me you are a leather smith. I can assure you that you will have plenty of challenge for your skills in taking care of the six horses and tack on the property.

"But let me give you one more reason why I hope you will respond favorably to my invitation, then I will give you as much time as you need to ask me questions. Will that be all right with you?"

"Of course," Avraham says.

"My last reason is just as simple, though of more of a political bent. You see, Avraham, I see a worsening of tensions between the Romans and the Jews in the coming years, and these can lead to war. Do not misunderstand. I am a senior officer in a Roman legion. If my legion is assigned to go to war, then I will go, and will command my legionnaires in ways that virtually guarantee we will win any battle. However, in peacetime, I will do all in my power to avoid war, and I will do all in my power to root out the seeds of war if I find that they will grow if left unchecked.

"So, my main reason is one of diplomacy. I would like to do everything I can to lessen these tensions so that we may both continue to prosper. For diplomacy to flourish, I need a much deeper understanding of the Jewish people, and I believe you and your family, over some period of time, can give me that."

He looks hopefully toward Avraham. For his part, Avraham looks at the space between them and says nothing. Quintus waits.

"I don't disagree with your view of the tensions between Romans and Jews," Avraham says at last, "but beyond the travel and the opportunity to see Rome as something other than a slave or prisoner, I don't see much of lasting value to me or my family if I were to accept your offer. However, I do see a great risk. What if you change your mind? What if you decide to keep us as slaves, or to sell us to someone who does not feel as kindly as you do toward the Jews? Perhaps of most importance to me is what will happen to my four children. They are good children, and they are too bright and inquisitive to be happy peeling fruit and emptying chamber pots at the direction of your wife, no matter how pleasant she may be."

Quintus pushes the loaf of bread he has been dipping toward Avraham. "Rest assured that I will be paying you and the members of your family for the work you do for our estate. Your work with the horses will be well known among other horse owners in my area, and I will encourage those needing your skills to do business directly with you. Though I will want you to oversee the planting, cultivation, and harvest, I also expect you to hire the members of your family and to recruit others of your choosing when necessary to assist with the farm work. "I am aware, in particular, of your two daughters, Gila and Ziva. My decanus has told me of their boisterous behavior in the marketplace of late. They will both be in the care of my most trusted household staff, Kema, an Egyptian slave woman. Kema is with me here in Jerusalem. She is looking after my household on the Mount of Olives. I will make sure you have an opportunity to meet her before much time passes. She is an educated and intelligent woman. She will keep your daughters challenged in all that they do, including the sciences, the arts, and mathematics."

Avraham folds his arms. "I also have two sons. Elias and Avi. Only Elias is old enough to work."

Quintus nods. "Your son Elias will work with another Egyptian slave on my staff. His name is Anyim. He, too, is very intelligent and is educating himself in his spare time when he is not repairing and replacing all the items necessary for the proper functioning of the house and properties. I am certain that Anyim will welcome some help from Elias."

Avraham is thoughtful again. He pushes the bread back toward Quintus without taking any. After a few moments, he speaks. "You have given me many generous guarantees. I know your title and status in the structure of the legion, but I don't know your real status within the Roman army here in Judea, let alone your relationship to the emperor in Rome. If you lose their support while my family and I are under your personal protection, I fear the worst for us."

"I think I can arrange a secure way to get your family back to Jerusalem without my protection," Quintus says. "Let us consider those possibilities as part of our larger negotiation. For now, I think we have laid out our concerns, and you need to begin a discussion with your wife. I would urge you to keep this entire discussion between you, me, and your wife for now." He peers down at the untouched loaf of bread. "Is there some reason you don't wish to break bread with me?"

"Only that Passover forbids my eating of leavened bread. Nothing more." Avraham leans forward. "There is one more important topic I want to discuss. You and your position in the legion are well known in the markets here. If I return to Rome with you in apparent friendship and collegiality, I may never be able to return to Jerusalem with the degree of respect among my peers that you claim you can guarantee. In fact, by the time of my return, any form of collaboration with Rome may condemn me, as it would with any Jew, to permanent exile or worse... to my death."

Quintus nods. "Any guarantee among men is subject to the whim of the gods—your one god and my many. The trip from Caesarea to Rome by ship will probably take most of a month if we have good weather, fair winds, and strong oarsmen, but my point is that there is no guarantee against powerful winds at any time on the northern stretches of the sea.

"The gods are always willing to participate in the affairs of men, and when they do participate, the first to go on the chopping block are often the guarantees made in good faith between men. The thing to keep in mind is that the gods always need to be kept well fed, happy, and unconfused."

He stands and dusts off his uniform. "I will leave you now. If you are contacted by Decanus Kanutus in the next several days, please pay careful attention to everything he has to tell you. Good day."

With that, Quintus departs. Avraham wonders if he should have stood and shown some kind of respectful acknowledgement of Quintus' status relative to his own.

'Too late now,' he thinks to himself. In the meantime, he decides to return to leather market, and to figure out how to talk to Orli about all that he has heard.

***

Ultimately, after many hushed conversations with Orli, late at night at Aharon's kitchen table, Avraham and his family go with Quintus to Rome. First, they returned to Bethlehem after Passover. In early summer, neighbors reported that they had disappeared from the streets there in the way that many Jews had disappeared from the streets of Jerusalem following Herod Antipas' instructions regarding the Jews and the Roman gods—quietly, in the night, and without a word to friends or associates.

So far as anybody knows, even close family friends, Avraham and his family have been taken by the Romans. As with many other cases, anyone who asks the Romans where Avraham and his family might have gone off to is met with silence.

On being told of their disappearance, a common response became: "Ach. I knew Avraham's making fun of the Roman gods would someday ruin him and his family. I am so sad for the children."

Gila would tend to agree. She is sad for herself. In fact, she is miserable. Her father has explained their situation to her and to her siblings, but she still does not feel safe. Her friends in the marketplace said that all Romans are liars. Their fickle promises can't be trusted, and besides, she does not wish to serve them.

As the shores of her home drift farther away, she wallows in feelings of self-pity. She had always known that attending Passover in Jerusalem next year was never very likely, but now it is impossible. She knows for sure that she won't see the young Nazarene man, won't watch him perform his miracles, and this makes her glum. She and Ziva try to boost their spirits by playing a game of hide and seek.

It is while shirking her duties, hiding from her sister in a mostly empty grain barrel, that Gila hears footsteps approaching. Even more than she does not want her sister to find her, she does not want an adult to fuss at her for her insolence, so she burrows down deeper into the grain.

"I'm sorry, Quintus," she hears her father say, "but even with your full assurance, I am still concerned about your frequent absences from Rome on military business. What if you get called to duty in some remote province and get killed or wounded there? I'm sorry to bring such a ghastly image to mind, but how will we get home?"

Gila's little forehead puckers. She hadn't thought of that. She listens closely to Quintus' reply.

"Avraham, please. I have promised to make arrangements if such a horrible event came to pass, but, if it makes you feel better, I will ensure that Decanus Kanutus and I show you around Rome and introduce you to our particular set of friends and associates, who will make sure no harm be-falls you, your wife, or your children. Can you agree to our assurances?"

Her father still sounds conflicted. "Too many possibilities..." he mutters. The legionnaire must give him a stern look because his voice changes. "You must forgive me, Quintus Caecillius. I am a man grown skeptical by my life on this earth. It will take time for me to fully invest my faith in you, but I do not doubt your intentions."

"I have asked you for a great deal," Quintus says. "That is all I can expect for now."

The footsteps fade, and Gila emerges from the barrel. She takes her father's lead: she is not sure what to feel in light of Quintus' words, but for the time being, she won't stand fully on faith until the centurion has earned it.

On the sea voyage between Caesarea Maritima and Rome, the weather is fairly calm except for some contrary north winds near the Greek port of Rhodes. By his status in the legion, Quintus is able to command space on a military trireme with a larger sail, and a keel and a hull that have been lengthened to include more rowers and lightened by removing much of the battle gear to enable more speed.

Though Quintus and his precious cargo are not trying for speed, his military attitudes as a centurion—and, therefore, the superior officer on the vessel—are always in play. He keeps the pressure on the oarsmen for a steady and deliberate rowing pace during windless days on their journey toward Rome.

Along with Gila and her family, Quintus travels with his housekeeper in Rome, Kema, his Decanus, Kanutus, and two of Kanutus' most trusted subordinates.

Avraham's family, always mindful to keep up the charade of their slavery, pass their time on the trip helping with cooking, cleaning, and loading supplies when in port. Elias is particularly eager to help in running messages between Quintus and the captain, and then the captain and the rowing supervisors and sail pullers. When not running messages, he is ready with his full water bucket to go up and down the ranks of rowers several times a day, always eager to satisfy the thirst of the sweating oars-men.

Most of the rowers are very grateful for the attentions of the children of Avraham, though their patience is tried by the constant questions from Ziva and Gila. They want to know everything about the oarsmen's lives when they are not on the vessel. Kema keeps careful watch over the girls when they go on these forays among the rowers. Kema is quick to gather them up and carry them away at the first sign of a scowl on a tired rower.

***

It seems that an eternity has passed since Gila boarded the trireme with her family. Finally, though, they make it to the port of Ostia outside of Rome.

Quintus' estate is near the small community of Vitinia, a few miles west of Rome. Quintus has been as good as his word that Avraham's family would be well received by his household staff. His wife Cornelia, the overseer of the estate in his absence, goes out of her way to welcome the newcomers and make sure they are comfortable in their new home.

Kema introduces Avraham's children—from the youngest, Avi, all the way up to the oldest, Elias—to the household staff. Gila and

Ziva become quick friends with Shani, an African girl who works directly for Cornelia; and Anyim, a young African who works with Quintus when he is on the property and with Cornelia when Quintus is not available.

After the introductions, Quintus motions for Avraham, Anyim, and Elias to follow him out to the stables. Quintus introduces Avraham and Elias to the six horses, and to the arrangement of the tack and harnesses, the workshop, and the tools. Avraham can see that he will have many happy days working in this fine shop. These good beginnings continued with only minor conflicts and upsets for several years. Ziva and Gila help all the members of the household staff with their conversational Greek and Hebrew, the conversational language of Jerusalem, and Orli works with Cornelia on the general oversight of the sometimes rambunctious children and adults. When they have time available, the male members of the household, including Quintus and Kanutus, also find themselves being tutored in Hebrew language and reading by the girls.

Gila was greatly pleased, and her curiosity greatly rewarded when she discovered the library in a corner room of the house. She had learned to read early, whenever someone would teach her, but she knew that her ability was rudimentary at best. This library, and Cornelia's promise to help her read any of the scrolls in it, left her awestruck. Out of politeness toward the centurion's wife, she expressed her appreciation quietly. Secretly, though, she was barely able to contain her impulse to pull a scroll off the shelf and begin reading at once.

Ziva is also a reader, though somewhat less intense about it than her sister. Ziva's enthusiasms are more for writing and the use of calligraphy in forming the letters of the Greek, Roman, and Hebrew alphabets. The library contains so many scrolls and maps that Ziva immediately realized that she could not master all of the writing styles no matter how long she remained in the Caecillius household.

Kema, Shani, and Anyim also prospered by their close proximity to Avraham's family, particularly the children, and most particularly by the training in language conducted every day by Ziva and Gila. For the Africans training in everyday Greek was constant with reminders, penalties, or rewards as needed, and at any time of the day, and as determined by the girls.

Gila can be pushy and is often intolerant of errors. She is often short-tempered with whoever had allowed errors to happen. Anyim, growing quickly into manhood, often became frustrated by this abusive treatment by a child, and would stomp out of the room. After a cooling-off period, one or the other of the two combatants would extend a hand of friendship, and the learning would then resume.

In the summer of her sixteenth year Gila sits in the library, curled up in her favorite spot, on a bench covered in soft blankets, near a window facing south. She has learned the names of the places beyond it: Vitinia, Ostia, the wild lands gradually being forced into Roman civility beyond.

She is young, but she hates when her parents tell her so. She knows more than her sister and brothers, more even than many of the politicians Quintus brings to dinner, and she does not appreciate having to put on the role of slave girl for them.

On this day, she is researching the works of Cicero; his subject the Roman Senate. Her father has suggested that she might compare his thoughts to writings in the Torah, primarily ones about Sanhedrin tribunals. She is chewing the nib of a stylus, wondering how she might weave logic between the two works like the strands of a plait, when Quintus himself walks in. He smells strongly of sweat and hay and warm horse, so Gila surmises that he is stopping by on his way in from the stables.

"Salvete and well met, Gila," he says. "What are you learning today?"

"I'm exploring the similarities between Roman and Jewish governance," she says, hoping to get a rise out of her so-called master. She is of the age that such things please her.

"The similarities, you say?" He frowns as he takes a seat opposite her. He digs beneath the strap of his sandal to satisfy an itch. "And what might those be?"

She hadn't expected him to take the bait. She gnaws on her lip, doing her best to come up with an answer that he'll accept, that won't insult him or ruin her father's position in his house. "Well, sir," she begins, "I wager that they both operate on behalf of the people they represent. Your Ro-man senators and consuls make decisions based on what is best for their people, even electing the emperor at times, while the judges in our Sanhedrin use the word of Jehovah as a basis to settle everything from land disputes to exemptions from tithing." She scratches her nose. "Allegedly speaking, of course."

This amuses Quintus. "Of course. So, does Miss think that either governing body represents its people truly?"

Gila thinks of the young carpenter from Galilee, who had stood in the market and lambasted the Sanhedrin. Rebellion thrills down her spine at the memory. "Not always. I feel that they could be more democratic, have more care for the common man."

"Ah! How very Greek of you, like Cicero himself." Quintus knocks his signet ring against the table between them and stands up. "Keep up the good work, Gila. You have a strong mind, and it will serve you well."

Avraham enters the library. As Quintus departs, he and his daughter exchange warm glances as they bask in the glow of Quintus praise for Gila's initiative in her studies. As time passes, training in the spoken languages of the surrounding community are expanded to include lessons in writing for all of them by Ziva, and by special tutors brought in for the purpose by Cornelia.

Avraham maintains his part of the bargain with Quintus. Several times each week, and more frequently during important Jewish festivals, Avraham sits down with the centurion to discuss one of the many topics that he wants to know about. In return, Quintus takes Avraham with him when he goes into Rome on either legion or personal business. Quintus introduces Avraham to his professional friends and business associates as a slave, captured with his family in Jerusalem. After a pause, he always adds, "...and a very valued member of my household staff."

Avraham's work with the horses allow him to compete with other horse owners in his neighborhood and in Rome for selection to two chariot teams. Chariot races are held regularly in the Circus Maximus a little way south of the Roman Forum. These intense and exciting races provide the final test of a horse's ability to carry a chariot driver and a centurion, or a chariot driver and a pair of archers into battle.

This, in addition to the cursus publicus, is much of the purpose of raising horses on the rural estates outside the city. The legion pays well for good horses, and Quintus is able to earn significant amounts of money from the horse husbandry that Avraham brings to his estate.

***

Shortly after the Roman midwinter celebration of the solstice, Quintus calls his household together to discuss their return to Jerusalem.

"I have asked you all to join me to discuss a matter of considerable importance to all of us, though it is of particular importance to Avraham and his family. The governor of the province of Judea Palestine, Pontius Pilate, has asked me to gather my staff and others who can support my work, and come to Jerusalem to help resolve some emerging conflicts be-tween the Jews and the Romans. These conflicts need to be resolved soon because they have the capability of escalating into situations that can be violent, situations that none of us want to contemplate. Pilate and I want to do all that we can to bring about as many resolutions as we can."

Quintus pauses a moment, then turns to Avraham. "Since I first met Avraham in Jerusalem, we have discussed the day when he and his family must return to their homeland. That day is now rapidly coming, and it is time that we all hear his thoughts on the very important decisions that are coming upon us."

Gila is so thrilled by this revelation—home! —that she almost doesn't hear her father's question. "You have not said how many people will be allowed to accompany you," Avraham says.

Quintus responds. "Pilate sent his formal request to me by personal courier. The official document was signed with his own very distinctive signature. Along with Pilate's signature, there were several signatures of my colleagues in the Tenth Legion. I don't think anybody will refuse to honor whatever request I might make in terms of the people I chose to bring with me."

"So, you are saying that I can bring my entire family back to Jerusalem?"

"Yes. Though I think you need to discuss this with each of your family members before making your decision. Under the tutelage of my Cornelia, Orli has become a very accomplished fresco painter and sculptor; Elias is now a grown man and has gained many skills that are very useful, and valuable, in animal husbandry and maintenance of an estate of this size; Ziva's teaching skills in reading and writing are very impressive. These skills are valuable anywhere within the Roman Empire, and their language abilities, as well as their ability to read and write, mean that they can make their way anywhere. I would ask each of your children to con-sider whether they want to pursue their futures here, in the heart of the Roman Empire, or in the distant, rural provinces of Syria and Judea."

Orli speaks up. "My children, except for the youngest, Gila and Avi, can make up their own minds on their futures. I will go where my husband goes."

Quintus thinks to himself for a few minutes. He rubs his chin as he often does when confronted with a difficult and tangled problem. "I want to speak about Gila. She is now an almost grown woman, and a very attrac-tive one at that, but her real beauty is not only in her pretty face. It is in her mind. In addition to her constant work trying to bring all of us up to some unreachable standard of literacy, I have taken note of the time she spends in the library looking at the scrolls and making notes to herself.

"I hope everyone understands that I very much want Gila to come with us to Jerusalem. Her erudition and charm will work miracles on the contending parties there. I can't imagine trying to do my job without someone as talented and as competent, but it is up to Avraham and his family to make the decisions on which of his family, besides himself, will come with us."

Several similar meetings take place over the next few months as each member of the household makes their personal and collaborative deci-sions on whether to stay in Rome or move on to Jerusalem. Finally, it is agreed that Avraham will take Elias and Gila with him. Despite her desire to go with her husband, Orli decides – after extensive discussions with Avraham - to stay on the estate until Avraham calls for her to return. Neither she nor young Avi will have much to contribute to Quintus' mission. At the same time, she has much to learn about the visual arts and the wonderful and colorful clays and stones that she has become so adept at working into the creation of beautiful objects.

One of Orli's most prized works is fashioned around a thumb-sized piece of almost translucent blue polished stone. She has crafted a beautiful flowered medallion from gold to hold the stone and with Cornelia's blessing presented it to Quintus. On receiving the beautiful piece Quintus says nothing for a moment; then, his eyes red with tears, he says that he had not thought about the stone for many years, but he could not have imagined a more beautiful gift nor could he imagine a finer piece of craftsmanship.

Gila is not sure why the Roman is overcome with emotion upon receiving her mother's gift, but she imagines it has some personal value to him. At any rate, the setting is breathtaking. She runs to her mother and gives her a great hug. " Mother, your work is so beautiful," she says.

Orli breathes into her daughter's neck, inhaling the smell of a girl teetering on the precipice of womanhood. "Thank you, haim shelli," she responds.

***

The sea voyage is mostly without incident except for some stormy weather and stiff, choppy, northerly winds in the crossing from Corinth to Rhodes. Pilate is eager enough to have Quintus in Jerusalem that he has yet again requested one of the faster triremes to transport him.

As the sea wind whips Gila's hair, she idly wonders if she'll see Jesus there, in Jerusalem. She wonders if she'll appreciate what he has to say even more, now that she's grown in wisdom and in stature from her tutors in Quintus' library.

Even with a strong crew, the choppy winds in the Aegean cause a lot of rocking. Rowing becomes difficult. Using the sail to power the ship becomes difficult and dangerous on the surging deck. The guests of honor on the boat – Quintus, Decanus Kanutus and his two Legionnaires, Avraham, Kema, Gila, and Elias – are advised to stay in their quarters below decks at the rear of the ship for their own safety. Initially, Gila and Elias are both disappointed that they cannot mingle and work with the rowing and sailing crew as they had done when Quintus first brought them out to Rome from Judea. After a few days in enclosed spaces on rough seas, though, the entire party, even Quintus and Avraham, are glad that they have no immediate responsibilities.

Quintus is the ranking officer on the trireme, but the ship's captain knows that he has a free hand to deal with the weather and the crew how-ever he sees fit.

As always Quintus is inclined to allow the captain full authority over the trip, except when the captain has to recommend that the ship be taken to an area of calm where repairs can be made, and the crew given a period to recuperate. Quintus knows that Pilate is eager for his arrival, and any unnecessary delay will not go well for either the captain or himself. He al-lows the captain one such unscheduled stop for crew rest and ship repair at the Isle of Rhodes, a port on the eastern end of the Aegean, just south of Rome's Asian province.

Fortunately, the winds to the east of Rhodes abate, and the remainder of the trip is relatively calm.

***

At Caesarea, Quintus and his party are met by a welcome party sent by Pilate. To Quintus, the welcome seems warm and genuine enough, although the delegation spokesman, a tribune named Ephesius, suggests that they take a few days to get rested and settled, then plan to meet with Prefect Pilate midmorning on Friday.

"The scale of events is growing more intense. Passover is coming soon. The contending parties need attention, and we need some of your skills in diplomacy. Are you and your staff up to it?" he asks.

"I have every confidence in my staff. If you will give us a chance to get our feet on the ground and get cleaned up and organized, we will be ready to discuss our plan with the prefect," Quintus replies.

Three days later, Quintus and his private staff of Avraham, Gila, and Elias arrive at the governor's offices to meet with Prefect Pilate. Decanus Kanutus and his legionnaires have been detailed to acquaint themselves with the senior legionnaires in order to give

Quintus a report on the likelihood of receiving their full cooperation if and when it's needed.

Quintus notes that, as usual, Pilate likes to keep important visitors waiting as a reminder of his authority over the agenda and schedule of whatever meeting is to occur. Senior legionnaires like Quintus recognize the need to tolerate these delays even when there is urgency in the topic; Pilate has a nasty habit of abruptly cancelling important meetings if he becomes offended by the actions of one of the participants. Sometimes he does not reschedule the meetings. In this case, however, Quintus is sure that Pilate wants this meeting more than he does.

Eventually, Pilate welcomes them in. After an exchange of greetings, Quintus, Avraham, Gila, and Elias sit down on one side of the table, and Pilate and Tribune Ephesius sit down opposite them.

"Two of your staff members seem very young, Quintus," Pilate says.

"Do they have the skills and experience necessary for such an important mission?"

Quintus replies, "The two youthful members of my staff, along with their father, Avraham, have been slaves in my household since their capture in Jerusalem nearly eight years ago. They have lived and worked with my family in complete harmony for that whole time. They have immersed themselves in Roman culture and attitudes. They have learned and have become adept at all the languages of the Roman Forum and marketplaces. At my request they will be training selected members of my legion staff in the nuances of Greek and Hebrew in the Jerusalem marketplace.

"As you know, I have an extensive library of writings by prominent Roman, Greek, and Egyptian writers. Gila, in particular, is now fluent in those languages, and she has become learned in many of our most important writers. She has been developing a comparison of the Roman gods, with references to the Jewish

Torah, based on her research in my library. She is young, but I believe she can hold her own in any discussion of nuance between Roman and Jewish philosophers."

Pilate looks at Gila, then, after a pause, looks again at Quintus. Pilate's reputation for lechery is on Quintus' mind as he waits for the prefect's next question. "So, Quintus. A very solid endorsement of this very beautiful young woman. Avraham and Elias? How do you see their role in any mission I may assign to you?"

"Avraham has been taking care of my horses," Quintus answers for them. "The horses are bred for sale to the legion. Avraham has developed a very fine breeding line of such horses. He works with other horse wranglers in my area west of Rome to develop racing horses for contests in the Circus Maximus. The horses bred under Avraham's husbandry consistently win those contests. I believe that whenever he is called upon to share his knowledge either with your horse handlers, or with any other Roman horsemen or handlers we might come in contact with, the dealing will have very favorable results."

"And Elias. What special talent does he bring?"

"Elias has been responsible for the day-to-day workings of my estate, and for making sure that everything is in good working order, so it is always ready when needed. He works closely with my wife Cornelia, and with my household staff in taking care of purchasing and storing items needed by the household. He works with his father in taking care of the horses, and with his sister, Gila, in helping with the language education of all the members of the household.

When I send either of them into the marketplaces of Jerusalem or Rome, I trust absolutely in all they will report to me on their return. I consider them to be indispensable," Quintus concluded.

"There is one last note that I want to pass along to you," Quintus says to Pontius Pilate. "Avraham, Gila, and Elias are Jews. They were born Jews and have been Jews all their lives. I have not interrupted their study of their books, and I have not prevented them from attending synagogue in Ostia. This is not favoritism on my part. My two other household slaves are Nubian from the mountains at the sources of the Nile river near Aethiopia. I do not interfere with the practice of their religion either.

"For myself, my wife, my children, my household staff, and the legionnaires on my personal staff as centurion primus, we are Romans, and we live by the Roman gods. I encourage all the members of my household to observe as we pay obeisance to our gods, which they sometimes do, and they invite us to return the favor, which we also sometimes do."

With a barely concealed sneer, Pilate interrupts. "That is all very noble and generous of you, Quintus. You have convinced me of the qualifications of the people you have brought with you for this assignment. To have brought such accomplished and knowledgeable Jews will no doubt be of interest to Herod Antipas, the tetrarch, and may help gain his support for whatever you propose to do. But my patience grows thin, the day grows long, and my reasons for bringing you to Jerusalem have not yet been introduced."

He sits forward. "The Jewish Passover will be upon us in about six weeks. It is important that you and your staff be there for the weeklong celebration. Over the past few years, there has been a growing stir east of here, in the cities of Sepphoris and Tiberias, and in the farming areas and villages around the Sea of Galilee.

"This particular stir—and the reason I brought you here—concerns a new, badly dressed prophet of baptism, and a new movement that has grown up around his preaching. While I am usually not much concerned about these minor sects that grow, live, and soon die outside Jerusalem, I am concerned when the Jewish establishment in Jerusalem gets all excited and begins to describe one of them as the new Messiah.

"Herod Antipas, has had the most prominent of the preachers, John the Baptist, arrested, but not before he baptized one of the other well-known preachers out there, Jesus of Nazareth."

At the mention of Jesus of Nazareth, Gila can feel her eyebrows start to furl. She quickly restores her face to the submissive state she is trying to maintain for this meeting. Fortunately, Pilate is turned toward Quintus, and does not see the change. Quintus, however, takes note of the subtle shift in her demeanor.

"The legionnaires in these rural areas have told Herod that an arrest of another one of these baptized preachers, particularly the Nazarene, will generate a lot of anger among these Jews in Galilee," Pilate says. "This is a growing concern. The Jewish 'nobility,'" he says this with an arched eyebrow and rolled eyes "the Sadducees, Pharisees, and the Sanhedrin, are also concerned about this 'baptism' because the idea of total immersion as the only way to wash away sin and corruption does not conform to their most cherished beliefs. They are equally concerned about this Jesus fellow.

"So far, I have said that I have no plans to interfere with Jewish teaching or teachers, but I cannot speak for Herod, or for Phillip, the tetrarch of Gallica and Syria. If the Jewish leaders are concerned about the welfare of John the Baptist while he is being held captive, I have told them that I will carry a message to Herod. I will recommend that he be released to your custody Quintus. If that is not possible, I will strongly recommend that they make sure he does not come to harm while in their custody.

"To me it is either make all the religious sects in the provinces of Syria, Gallica, and much of Judea angry as hornets, or make a small and clubby group of Jews who believe themselves to be infallible understand that they have no monopoly over the practice of faith, especially when it is their own faith they are worried about. here, in the cities of Sepphoris and Tiberias, and in the farming areas and villages around the Sea of Galilee.

"This particular stir—and the reason I brought you here—concerns a new, badly dressed prophet of baptism, and a new movement that has grown up around his preaching. While I am usually not much concerned about these minor sects that grow, live, and soon die outside Jerusalem, I am concerned when the Jewish establishment in Jerusalem gets all excited and begins to describe one of them as the new Messiah.

"Herod Antipas, has had the most prominent of the preachers, John the Baptist, arrested, but not before he baptized one of the other well-known preachers out there, Jesus of Nazareth."

At the mention of Jesus of Nazareth, Gila can feel her eyebrows start to furl. She quickly restores her face to the submissive state she is trying to maintain for this meeting. Fortunately, Pilate is turned toward Quintus, and does not see the change. Quintus, however, takes note of the subtle shift in her demeanor.

"The legionnaires in these rural areas have told Herod that an arrest of another one of these baptized preachers, particularly the Nazarene, will generate a lot of anger among these Jews in Galilee," Pilate says. "This

is a growing concern. The Jewish 'nobility,'" he says this with an arched eyebrow and rolled eyes "the Sadducees, Pharisees, and the Sanhedrin, are also concerned about this 'baptism' because the idea of total immersion as the only way to wash away sin and corruption does not conform to their most cherished beliefs. They are equally concerned about this Jesus fellow.

"So far, I have said that I have no plans to interfere with Jewish teaching or teachers, but I cannot speak for Herod, or for Phillip, the tetrarch of Gallica and Syria. If the Jewish leaders are concerned about the welfare of John the Baptist while he is being held captive, I have told them that I will carry a message to Herod. I will recommend that he be released to your custody Quintus. If that is not possible, I will strongly recommend that they make sure he does not come to harm while in their custody.

"To me it is either make all the religious sects in the provinces of Syria, Gallica, and much of Judea angry as hornets, or make a small and clubby group of Jews who believe themselves to be infallible understand that they have no monopoly over the practice of faith, especially when it is their own faith they are worried about.

All of Quintus' party rose, gave a proper salute, and departed.

***

As part of their education process, Avraham, Elias, and Gila have come down to Jerusalem from Caesarea to reconnect with their friends from years ago, Bathsheva and Aharon. Avraham hopes that Bathsheva and Aharon will agree to invite Centurion Quintus along to the Seder, and to other celebrations and rituals that will take place during the week of Passover.

Avraham explains their new situation and what they are now doing on behalf of the Centurion Quintus and the Roman Prefect Pontius Pilate. He explains that Pilate himself has brought Quintus back from Rome, and that Quintus has brought Avraham and his family along as indispensable staff.

Avraham has complete trust in Bathsheva and Aharon, but he knows that the issues and problems he is telling them now will test that trust in ways that none of them would ever have anticipated in their many years of friendship.

"You must trust me, Aharon, when I tell you both that Quintus bears no ill will against any of us here, or against any of the Jewish people. Based on our family's several years of intimate service in his household, I believe he and the prefect are working toward a long-lasting peace that will include all of Syria, Palestine, Gallica, and Judea. This is in the long-term interest of all Jews and all Romans."

"I would trust you with my life, Avraham," Aharon says.

Bathsheva also spoke: "I share my husband's feelings, Avraham. I only wish your entire family were here with us. I am overjoyed to see what a handsome and intelligent young man and woman Gila and Elias have become. If possible, can you have Quintus visit with us before the start of Passover so we will have a better idea how to introduce him to our friends at the various events? As I am sure you can imagine, there will be many questions. We must be open to whatever is said, and we must be willing to fully answer whatever questions are asked.

"I'm sure you realize, Avraham, that there is an increased level of tension in Jerusalem because of the capture of John the Baptist by Herod Antipas several months ago. The longer he remains in prison, the more this tension increases the risk of violence during Passover. I am also told that the Nazarene, the most prominent of those Baptized by John the Baptist, will be coming to Jerusalem from the Galilee, and may already be here."

Gila cannot contain her excitement at this. "Oh, Father! I must go to meet him! I know him, and I am sure he will remember me from the meeting Ziva, and I had with him on the streets here several years ago." Avraham smiles as he puts his hand on Gila's shoulder. "I don't know if I have told you that Quintus and I have been discussing such a meeting, Gila. It is potentially fraught with difficulty and risk. Jesus is probably

a much different person than the one you met. The cult of baptism is a challenge to the established rules of worship by the Sadducees and Pharisees, and they are very sensitive to the possibility of further degradation of their authority.

"Depending on the setting and circumstances the Nazarene is the kind of speaker who can turn a simple discussion of faith into an incitement to violence. Both Quintus and I are very concerned about this. It is contrary to what Quintus and Pilate are trying to do here, and, more than any oth-er consideration, I fear for your safety."

"Do you trust me father?" Gila asked.

Avraham is surprised at this bold question, but answers "Yes. Of course, Gila."

"Then let me work out a plan for how such a meeting might take place with safety, and how it can be made to serve the needs of all of us. Elias, will you help me? Will you agree to go with me to whatever meeting I am able to work out with the Nazarene?"

"Of course, my sister. I will follow you wherever you go, and I will protect you with my life."

Avraham and their hosts seem to be struck silent by this expression of confidence and faith by these two children. Avraham knows that he must say yes to Gila's request.

"Can you be ready to discuss your plan with Quintus and his legionnaires in three days?" he asks.

"Yes, Father."

Avraham rises to gather his family and thanks his hosts for their hospitality. He asks, softly, if they might be willing to work with Gila depending on what she may propose.

"Yes, Avraham," says Aharon. "Of course, you may count on both of us to do all we can in the interest of lasting peace." Bathsheva nods her assent. She looks deeply into their eyes and smiles when each of them looks directly at her in turn. Gila proposed a meeting with the Nazarene close to one of the main streets in the lower marketplace, but at least partially isolated from the crowd that would flood Jerusalem on this precious and popular holiday.

Quintus consented to these arrangements. He would attend the meeting with Gila. They would invite the Nazarene to bring along whomever he might choose. Discussions would take place in the language the Nazarene is most comfortable with. If the chosen language is Aramaic, Quintus and Gila will bring an interpreter skilled in the nuance of that language. If the Nazarene prefers a language he is not totally comfortable with he may bring also his choice of interpreter to assist him.

The purpose of the meeting is simply to explore issues and problems of concern to the Nazarene, with the promise that Quintus and Gila will carry his concerns to the appropriate senior Roman and Jewish officials.

Other meetings with the Nazarene will follow. Quintus will have a small contingent of no more than ten battle-experienced legionnaires with him. They will be in uniform with minimal battle gear. They will be moving casually, but together, and out of marching order in full view of passersby in the market. They will be ready to act quickly, and with deadly force if necessary. They will be trusted men who fully understand and support this initiative. They will give their lives to it if that is what is required.

***

Jesus the Nazarene agreed to all these proposed arrangements when he attended a dinner at the home of Bathsheva and Aharon a few days before the Passover Seder. Bathsheva and Aharon were both held in high regard by most of the Jewish factions in Jerusalem. In large part their good standing—and their success in persuading the Nazarene to come to their house—was a product of their good works in the Jewish community and their long association with Avraham, who had spent so many years work-ing in the hotbed of religious tumult around the Sea of Galilee.

When introduced to Gila, the Nazarene is very pleased to make her acquaintance again after many years. "I knew we would meet again, though I did not think it would be so long. I hope our time together will be profitable for all parties. With your full participation, I have no doubt that it will be." As the meeting breaks up, the Nazarene asks Avraham if he can have a few private minutes with Gila. Avraham agrees.

In the shadows outside the house of Aharon, Jesus speaks quietly in Aramaic: "I cannot tell you how pleased I am that you are involved in this, Gila. I knew when I first saw you that you had a great intelligence that will be essential as we Jews continue to work toward religious and cultural independence and security within the Roman Empire.

"But you must be careful. I'm sure you know by now that the leadership of the Romans and the Jews have a great capacity for deceit; they may say many things for effect, but they have very simple goals: the Pharisees and the Sadducees do not want to see the Baptists gain influence because they are a clear threat to the religious authority of the Sanhedrin. The Romans probably don't care how the Jews sort themselves out, but they have no interest in losing control of the situation in Judea, Syria, and Palestine. The Romans will sacrifice whoever is necessary on whichever side to insure that result.

"I do not want you to be hurt because I believe you, and Elias, and your family, and even Quintus and his family are very much at risk in this process that Pilate has set up, and I believe you are all very important to the future of this part of the world."

"Your words take my breath away," Gila says. "I feel a great burden, and I am frightened that I will not be able to do what I am being called upon to do. Can you help me?"

"I will be with you whenever you and your friends and family are among my people. For various reasons, I cannot be with you when you are among the ones who fear me, nor can I be with you among those who have called me 'enemy.' In those times you must rely on the word of Quintus that he and his chosen legionnaires will protect you from physical harm and will protect all of those you hold dear. I have looked into his eyes on many occasions, and I believe him to be a good man."

"How do you know Quintus?" Gila asks. "Have you ever met him?"

"The same way I know you." Jesus pauses. "I must ask you to do one last thing for me."

Gila finds that she has lost her breath. "Of course. Anything you ask."

"I have come to believe in the power of fully immersing the body in the baptismal pool. It is only in this washing away of all of the sins and corruption of the body that redemption and rebirth are possible. This is what I have learned from my own baptism by the hand of John the Baptist. I believe this, and I believe that all who have followed the example of myself and John also believe it."

She begins to feel a little uncomfortable. What is this leading up to?

"Do not be disturbed, Gila. I know you and all your family are devout Jews, and you have now become a very learned young woman. I only want you to realize that there are new ways of expressing devotion to God, and I want you to be aware of the one way that I am most pleased with. If you ever decide to be baptized, please be assured that I will want to conduct the ceremony. Will you keep me in your thoughts?"

Though she knows it is wrong according to the powers that be, though she isn't entirely sure what she is feeling, Gila leaps into the arms of Jesus and hugs him very tightly. She has not had these kinds of feelings before, and though they immediately make sense, she is still a little shocked and amazed at them, and what they made her do. For his part, Jesus hugs her in return with some of the passion he feels toward her. Their cheeks touch and rub together for a moment. Though he knows he is not above the sins of the flesh, he has no desire to take advantage of this beautiful child.

He turns her face toward his and kisses her fully on the lips. As the kiss ends, she pulls slowly and carefully away from him, unwilling to look away from the eyes of this man. Jesus smiles, then turns away and is gone into the night.

***

Pilate's plan has worked, at least for the time being. Passover goes with-out incident, and Jesus of Nazareth is able to move freely among all those who observe Passover with varying degrees of devotion and emotional intensity. As promised, Quintus has taken the concerns of Jesus to Pilate, and Pilate has given every assurance to Quintus that he will take them up with the tetrarch and with the other powers that be in the empire just as soon as it is appropriate.

The party that traveled to Judea with Quintus all return to Rome. Avraham discovers that his work with the horses in Quintus' stable is far more interesting to him than anything that might happen during the years left in his life in Jerusalem. Elias also feels that his future lies more in Rome than in Jerusalem. Gila returns to her now well-trodden path of scholarship and diplomacy. Quintus makes sure that she is enrolled in the finest teaching institutions available in Rome, and that she is given full access to all of Rome's philosophers and statesmen. Following the period of study in Rome Quintus has also made sure that she will have many opportunities to travel to the far corners of the empire. In this way Quintus believes she will learn the local languages and customs necessary to the completion of her training.

In Quintus plan for her she will then be ready for appointment as a high-level diplomat with a specialty in the religions and languages of Rome's eastern provinces.

As Gila prepares for a course of learning for several months at the Great Library of Alexandria in Egypt, Quintus approaches her. "I want to give this to you," the centurion said. From the pocket of his cloak he produces the beautiful blue stone, nestled safely in the gold work her mother had crafted. "For your work on our mission of peace in Jerusalem and across the Empire," he said, handing it to her. "My great-grandfather was given this as a commendation by Julius Caesar himself, in gratitude for his action on the field of battle against the last of the Ptolemaic dynasty in Egypt.

"I now give it in gratitude to the sacrifice you and your family made all those years ago. Like my great-grandfather, you are a warrior, Gila.

I know it may not have been easy to leave your home and all you knew, especially not for the little girl you once were, but I am very glad to have had your help."

For her part, Gila is pleased with all of this. She continues to train for and hone the diplomatic skills that she feels blessed with. She hopes that she will eventually reach the upper echelons of Roman society, but her mind does often wander to the night in Jerusalem when she was kissed by a man considered by most to be the Jewish Messiah.

END

COLLAPSE

Summer, 1350 CE. Agna, a young woman, works as a helper in a small bakery near the docks of Hamburg, Germany. Suddenly, a cry of plague rings through the streets. She grabs her small bag of belongings, and her few precious bits of parchment, her pen, and a small bottle of ink for her attempts at writing. She tries to escape the Black Death, which is spreading from the docks thanks to a cargo ship carrying the bubonic plague.

====

In Agna's escape from her job as a baker's assistant, she has joined up with two other escaping travelers: Sophia, a Catholic nun, and Meshek, a Jewish leatherworker. These three strangers decide they must learn how to work and live together during their escape to the Harz Mountains one hundred fifty miles south of Hamburg.

Part 1.

My family lives on the Baltic seacoast three days' walk north and east from the village of Lübeck. I remember my mother's frequent call as I grew up: "Over here, Agna, now please. Help your father load those fish."

Lübeck has become the administrative center for the entire region along the coasts of the Baltic and North Seas. For over two hundred years, the commercial interests along the thousand miles of coastline between Bruges in Belgium and Gdansk, three hundred miles east and north of Lübeck, have been building up trading relationships. These relationships—called, informally, The Hansa—link commercial interests in London as well.

I am short and well-muscled from my family's hard work. We earn our way by fishing and by helping other local fishermen and farmers with the shipping and selling of their wares. Fishing has not always been in my blood, though. There are rumors among some members of my family, who do not usually talk about such things, that a castaway had washed ashore during a storm in the

Baltic Sea. This was several generations ago. The castaway was said to be Asian. According to family rumor, the castaway found a way to keep himself alive. He built a small settlement that eventually became known as RabanHaven—home of the raven, in other words. The survivor spoke an academic but clumsy form of High German. He was good with tools and learned how to work as an itinerant laborer. He could earn money helping local fishermen and the warehousemen and teamsters who carried the produce of the area, mostly herring, to market.

Those who might have been curious about the survivor's foreign features and his command of high German kept to themselves about it after seeing how hard he worked for them when they needed help. He eventually married a local German girl. They had many children and grandchildren, and some of these offspring married and had their own children in time.

I am one of these fifth-generation grandchildren. My parents took note shortly after my birth that I am slightly different from other German children. For those who look closely, my eyes are a bit almond-shaped, as are those of others in my family, but none are quite as pronounced as mine. My mother and father continue to hope that my eyes will not condemn me to becoming an outsider in whatever community I choose to live in.

My parents also hope that one other part of our family history will never become generally known: the survivor who founded our small community was a Mongol officer in the Army of Subodei Kahn. He was wounded and left for dead in the Battle of Leignitz over one hundred years ago.

Though born and raised by a loving family, I soon yearned to see the world and add to our trove of stories. I knew that I must leave them to see something of the world beyond RabanHaven.

***

Some distant part of my extended family lives in Lübeck, the administrative capital of the Hansa, so I endeavored to reach them. Lübeck is about three days' walking southwest of RabanHaven.

My distant cousins run a small rooming house for fishermen who spend the fishing seasons working the waters off the Baltic coast. Soon after I arrived, I was able to work with them to earn room and board. Eventually, though, I tired of the need to empty the bedpans and to accomplish other low and menial tasks in my cousins' hotel. I thought of other ways to earn a living but finding decent work in Lübeck is difficult for girls without connections. Opportunities for work are also limited for girls without the kinds of special skills important to the Hansa, things like reading and writing contracts, agreements and bookkeeping. My only other option would be to earn money by selling sex to men with enough money to make pregnancy, or the degradation of the act itself when done with strangers, worth the costs, but I am unwilling to make those sacrifices.

***

Upon hearing my plan, and my resolve to leave Lübeck, my relatives sympathized with me, though they still needed somebody to empty the pans so they would not have to pay somebody else to do it. They tried to discourage me in my plan, but eventually, I decided that I would leave without their permission, if necessary and walk the forty miles to Ham-burg. I then resolved to work out some kind of peaceful resolution of concerns among my relatives. They finally relented to some extent and even gave me some money and a basket of food for my journey.

I leave Lübeck early on a sunny day. I begin walking the forty miles along the crowded commercial road connecting Lübeck with the city of Luneburg near the Elbe, upriver from Hamburg. Many people are walking or pulling carts along the ancient, muddy road in both directions. Most are carrying goods for trading in their destination city: Lübeck for those traveling north, and Luneburg or Hamburg for those traveling south. Their way is complicated by the number of horse-drawn wagons carrying barrels of salted herring from the Baltic fisheries to the docks in Hamburg.

The Baltic herring fishery is very productive. My parents would say that the herring form the commercial basis for the creation of The Hansa itself. The herring is preserved in salt, which comes from mines in Luneburg. Whenever a herring wagon from Lübeck must compete for roadway with a salt wagon from Luneburg, traffic on the narrow path gets tangled as one tries to get past the other.

I know many of the teamsters on these wagons because my family helps them gut and salt the herring or pound iron into the shapes necessary to make reinforcing joints and steel wheel straps for their wagons. I do not want to show my face to some of them because they have not treated my family or me well in the past, and their actions toward me now that I am on my own might be even less desirable. I believe my most important possession is the dark hood and cloak that warms my body, hides my face, and hides my femininity from those who might mean me harm.

"Is that you, Agna?"

A man on a wagon passing in my direction calls my name. I think I recognize the voice, but I am afraid to turn to see who it might be.

He persists. "Agna, it's Kurt. What are you doing on this road?"

I look up to see Kurt, a friend from my village. I smile but am still un-sure how to answer his inquiry.

Kurt calls out to his teamster to pull the oxen and wagon to the side and stop. He jumps down from the wagon and comes over to greet me.

Both of us are too shy to hug or touch in public, but we do exchange smiles and warm glances with our greetings. Finally, Kurt asks again what I am doing traveling by myself along this road with its many thieves and dangerous men.

"I am walking to Hamburg. I have decided that I want to see if I can find better work, since there is not much work in Lübeck for young girls from the villages. I left the hotel where my relatives generously allowed me to work dumping shit for these past several months. I am very happy to see you, Kurt."

"I am happy to see you as well," Kurt responds. "But the road is dangerous, and you should not travel alone. Please come with me on the wagon. I can watch out for you, and I know some people in Hamburg who may be able to help you find work and housing."

This is the best prospect I've had in a while, so I climb up on the back of the wagon and sit next to Kurt on top of a barrel of fish. We talk about things that have happened to us since we last saw each other in RabanHaven many months ago.

The teamster lightly cracks his whip, and the ox team leans into the wooden shoulder harness and moves the wagon back onto the road.

Talking is hard. The road is bumpy and rutted by deep tracks in some places, and it is deep with mud in others. After a while any conversation seems like too much trouble. There is some woolen cloth in the front of the wagon. I retrieve the cloth, pull it around myself, and then lie down on a plank that Kurt has laid on the tops of two barrels for me. In spite of the rough ride, my exhaustion gets the best of me. I fall immediately asleep.

Later, I wake up to find Kurt shaking me. He is whispering hoarsely in my ear.

"Agna. Get up. Please get up!"

What is it, Kurt?" I mumble.

"Knights. They are up ahead searching the wagons for illegal goods." I jolt upright. The knights are of the Teutonic Order. My parents told me they are sometimes contracted by the business interests of the Hansa, not to mention the Holy Roman Empire, to make sure the commerce moving along this particular road is approved for transport.

"I need to know if you have papers that authorize you to be on the road," Kurt says. "If not, I can tell them you are a family friend from RabanHaven, and you work in Lübeck helping me organize these shipments of herring."

My head spins at this news. "I didn't know I needed papers to walk along the road. Will I be in trouble if I don't have papers?"

I've learned that the teamster's name is Aldrick. He turns to the back of the wagon, then pulls his hood aside to look at me. He is a much older man than I had thought. His face is wrinkled, and the eyelids and the bags under his reddened eyes actually droop. I recognize Aldrick as a steel smith in RabanHaven who has worked his trade there for most of my life. He has done work with my family several times. He has seemed to always be ancient over all the years that I have known him.

His voice is raspy as he speaks to me.

"It will go better for you if you don't say anything, miss, and you should try and keep your face hidden. These knights all had early family members killed or wounded by the Mongols in their last battle with them in Poland a hundred years ago. They have long memories, and they might start getting excited if they look too closely at your eyes."

I turn to Kurt. "Give me some dirt and grease from the bottom of the wagon around the axle. I'll smear it on my face and try to hide my eyes. I don't think a Teutonic knight will find a dirty little girl attractive enough to ask questions. If they do ask, I'll start to cry and wipe the tears away with my cloak. I'll look at you with a very sorrowful look. You can tell them I am a little thick and a little afraid of them. That should further discourage them from wanting to talk to me."

When Kurt returns my stare, I see that he is impressed with how much I've grown, how street smart I've become, since we last saw each other.

The two knights are on horseback, and they are inching closer all the time. They are wearing their white capes over white singlets with a black cross emblazoned on the front. Both are armed with long swords. Two squires carry out their orders by crawling into the wagons, asking for papers, and, sometimes, demanding that a case or carton be opened for inspection.

When they stop at Kurt's wagon, a squire asks him a few questions. He seems satisfied with Kurt's answers, but then asks about my relationship to him. I am looking down and away from the eyes of the knights and the squires. Kurt tells them why I am with him. The story is a good one because it is mostly true.

I glance up briefly when I hear the squire talking to Kurt about me, but I quickly look away. The squire asks no further questions. He jumps down from the wagon and walks alongside the knights to the next wagon in the line alongside the road.

I cannot help but be impressed with the knights' dress, their carriage, and their demeanor. They look too young to be much concerned with losing a distant relative to the Mongols so many years ago.

It is at this point that Aldrick turns again to speak to me in his gravelly voice.

"These knights are trained from birth to believe that they won a decisive victory at Leignitz in Poland, but the truth is that the knights' armies were destroyed by the Mongols. These knights claim that their victory

is the reason the Mongols retreated from Europe a few months later and never returned. The way I hear it, the Mongols left Europe because they got word that their Great Kahn, Obedai, had died in their capital at Karakorum. They all had to return to their capital in order to protect their political interests.

"If you take no other advice from me ever again, take this: Never discuss what I have just told you with anyone. If you ever do, and the word gets back to any of the knights, that knight and his family will search the earth for you. They will never stop, and when they find you, they will kill you in a very slow and painful way." Aldrick paused a moment. "...but probably not before they have tortured my name, as your source of the information, out of you."

I feel my body start to tense with fear and nervous energy. My eyes grow teary with these revelations from Aldrick. "If this is so, sir, why do you burden me with these horrible tales?"

My voice shakes with fear. My whole body will not stop shaking.

Aldrick answers. "You travel alone. I want you to know these things so you will be better able to survive any encounter with whatever may lay in your path. Your mother and father have always known of your desire to leave your village. They have always treated me fairly. They once asked me to look out for you."

"All the knights and their families are now part of the commercial interests of the Hansa," Kurt adds. "They are much more interested in making a few florins off people like me who haul their salt and their fish, and empty their chamber pots, than they are in seeking revenge over old grievances—that is, except for those grievances that arise over any threats to the honor of their families, or their family names. Those grievances never go away, and their desire to redress those grievances never fades."

Kurt motions the old teamster to move out. As we ride, I think about all the events of this day. After a period of time the rocking motion of the wagon over muddy, rutted ground puts me back to sleep.

PART 2.

On first arriving in Hamburg, I spend several days wandering the streets, trying to talk to people who know Kurt and who might help me find food and shelter as my small store of money and food is nearly gone. First, I must find a job so I can repay Kurt the few florins he loaned me to live on. After a few days, I find a temporary place to sleep where I can spend time looking for a good job, and begin, then, to think further about what I really want to do with my newfound freedom.

I know I have my imagination and, just as with the knights along the road, I know how to use it when there are obstacles to overcome. I also have some skills and a tolerance for hard work over long hours. I long ago lost any personal pride I might have about dirty jobs during my time growing up with my fishing family, and at the hotel in Lübeck.

I know how to decide which of a batch of freshly caught herring are worth saving and which should be thrown away. I know how to gut the fish that are worth saving, and I know how to separate and salt all the premium parts and the leftover parts of the fish that have value.

I have my own gutting knife and sharpening stone. My father gave them to me on my sixth birthday. They are precious to me, and I'm resolved to not lose them. I have other skills of the kind needed around boats and the people who work on boats. I have skills with woodworking, and with the repair of fishing nets and sails. I know how to raise a sail, and I know how to take it down to gain advantage with each shift in the wind, particularly in the shallow waters near the rough coasts of the Baltic.

***

Kurt helps me find work as a baker's helper. The shop is owned by a man named Cadell. His shop sells flour and bread to pursers on ships traveling the Elbe River between Hamburg and ports along the coast of the North Sea. When ships need to load or unload cargo and restock for their next trip, they tie up at the Hamburg docks near Cadell's shop. Cadell values my skills as a baker's helper, as well as my ability to speak more languages than the clumsy German of most of the young itinerants looking for work every day on the docks.

One day a sailor, Matius, comes into the bakery to buy a loaf of bread. Matius tells me he comes into the Port of Hamburg every few weeks. Cadell is away, arranging a shipment of grain for baking, so I am left in charge of the shop. Matius is a very handsome man, and I am pleased to sell him a loaf for a few pfennigs.

Over a period of months, we grow close. When Matius is in port and Cadell gives me leave, Matius and I walk around the city. We enjoy the sights, sounds, and smells together

One day Matius tells me that he and some of his shipmates have heard stories of plague from sailors working on the Mediterranean Sea. Matius becomes upset as he describes those who catch the plague and then die a horrible death.

"First they get sick. Then these horrible growths called buboes spring up under their arms and in their groins. If those grow in their lungs, they will surely die within hours. Sometimes death does not occur for a few days. Some will survive the ordeal, and a few will never get sick, even when exposed to those who have the disease.

"If our captain even hears a rumor of plague in a port, he will immediately untie from the dock and head to sea; whatever the weather; whether or not all the sailors are aboard; and whether or not the loading and unloading have been completed.

"Whenever you hear the word plague, you must immediately gather some food and water, and leave the city," he says. "Don't approach the man or woman who says the word seeking more information. They may kill you by breathing on you or touching you, or he or she may kill you out of fear that you might have plague. You must promise me that you will leave immediately." Matius looks deeply into my eyes.

"Yes, Matius, I will."

***

One day there are several boats tied up at the docks, and our ovens work constantly. Even so Cadell sends me to the boats to ask pursers about the bread and flour they might need for coming journeys.

When not actually in the shop selling or baking, I often accompany our shipment of baked goods to the boat to make sure the purser is satisfied with the order and that Cadell gets properly paid. The occasional purser is unsatisfied with the order, and uncompromising with my attempts to settle the debt. I have learned when to demand that the purser take me either to the captain or to the boat's owners, if the owners live in Ham-burg. At this point, there is a combination of charm, sexual innuendo, assertive talking, yelling, and foot stomping needed to resolve these kinds of commercial problems. My attempts to write are not pleasing to me, but I try to record at least the day and the name of the ship when these problems occur.

Once, when I grew too frustrated dealing with a ship's purser named Friedrich, I went to the rooming house where I knew Matius was staying. We did not yet know each other well, but I trusted Matius to do right by me, and he had the size and demeanor of a man who could be ready for a fight when necessary. Even better, though normally calm, Matius knew how to look the part of a very angry man.

Today Cadell is away dealing with a shipment of grain needed by our shop. He is not available for counsel about how to deal with the problem purser. It is all up to me.

Matius readily agrees to come with me to the docks. We arrive to find the purser back at his desk hunched over his paperwork. Matius motions for me to go back down the gangway to wait for him on the dock. Matius then barges back into the room without knocking. Rather than go down to the dock as Matius had asked, I stand just outside the door to the purser's spaces. When I look in, Matius has his back to me, but I can see the purser.

The purser looks up, his face suddenly pale. "What do you want?" he asks.

"I want the money you owe my friend for the bread and flour she delivered this morning," Matius replies.

The bread was stale, and the flour riven with weevils." "Show me the flour," Matius says. "Now."

"No" is the reply.

Matius picks the purser up by the collar of his shirt, carries him to the door, and through it to the rail outside. Before the purser can grasp what is going on, Matius tips him over the rail and dumps him into the water twenty feet below. Matius takes a moment to make sure the purser can swim, then turns to go find the ship's cook.

The cook had been watching as the purser went into the water. He moves to intercept Matius, then takes him to the captain who is passed out in his room from drinking too much beer. I wait just outside the door to the captain's quarters. The cook shakes the captain, hard, then kicks him. "Captain! Wake up! This man wants to get paid, and the purser is acting like a scheisskopf again."

The captain mutters something about the keys being in the desk drawer.

Matius later tells me that he hears the captain mumble something to the effect of "Pay the bastard, and then get him the fuck off my boat."

The cook looks at Matius and winks. "How much?" "Four florins," Matius replies.

The cook uses the captain's key to open a drawer. He pulls out a small leather bag and shakes out the four florins. "No hard feelings, I hope. There was nothing wrong with the bread or the flour. Our purser likes to think he is a tough guy who can bully people, especially pretty women like your girlfriend, so he can cheat them out of their money. Then he takes the money that should have gone to the supplier and stuffs it into his own pocket."

Matius, not knowing how to deal with this strange crew, nods, but does not change his expression.

"This garbage scow may want to return to these docks again. If I am still aboard, I hope we can do some more business. Once I talk about this incident with the ship's owners, I doubt if you will see either the captain or the purser again. Can we shake on it?" The cook extends his hand.

"Yes. There are no hard feelings. I'll let my friend know what you have told me, and that you have assured me that she and her baking company will be welcome to supply your ship in the future."

I look to Matius as he comes out the door. I had grown concerned when he threw the purser overboard, then disappeared inside the ship with the cook. I hug him and hold him tightly for some moments.

Here's your four florins," he says as we walk away from the dock.

***

Since the formation of the Hansa in the early part of the century, Ham-burg has become an important shipping port. It has been built to handle shipping between the farms, factories, and other producers of goods and produce from northwest Germany; goods that are held in Hamburg's many warehouses to exchange with goods and produce received from European markets scattered around the North Sea, the Baltic Sea, the channel between England and France, and the distant ports on the far seas beyond.

I know that plague is certainly no stranger to the city of Hamburg. As the rumors of plague in other cities grow numerous, I grow more concerned. Matius' admonishment rings in my ears.

Some of the shopkeepers have begun whispering in the nearby market. I know I will have to get away from the docks, soon even get away from Hamburg itself, if I am to survive the plague that many suspect is on the way and, now, very close to the city.

***

A few days after Matius' altercation with the purser, I wake to hear a man running down the street, screaming: "Plague! Plague! The plague is here! It comes on a boat drifting on the Elbe full of dead sailors! Flee! Flee!"

On hearing this I jerk up in my bed. The plague has come to the Hamburg docks. I gather some clothes, and what money and food I have on hand, and stuff them in a sack. I begin to walk to a place where I can take a water taxi across the Elbe. I hope that word has not yet got to the taxi stand, or the prices will rise above my ability to pay.

I am lucky. I find a water taximan whose deafness means he cannot hear the growing number of shouts on the street. He takes me across the river and does not charge me extra.

Like Lot, I begin to walk south toward the Harz mountains, over one hundred miles away, without looking back. In the mountains the cooler air might stop, or at least slow the spread of the plague before it can catch up to me. I have heard rumors that this is true from travelers who escaped plague in England, Ireland, and Scotland the year before.

I am not alone on the road. Others are walking, though some from the upper classes ride in horse-drawn carriages. I am still dressed in my flour-stained work clothes. Several others on the road are similarly dressed, as though they felt they had to escape the city as quickly as possible, grabbing only the few handfuls of personal items they could carry as they ran out of their homes and into the streets.

As the day unfolds, more people join us walking south. Many are already dragging their feet. I am now grateful for my upbringing and its constant demand that I do hard physical labor. I am also glad for the months in the bakery lifting and carrying heavy bags of flour. I resolve to walk as far as I can on this first day. When the time comes, I think I will go into the woods and try to sleep a little. I know, though, that the road will only grow more crowded with refugees from the plague, and I have no idea what might await me in the mountains. I know I can only sleep for a little while.

Some of the wealthier residents of Hamburg shout from their carriages for us to get out of the way, as though their passage is more important than the struggles of those of us carrying all of our belongings on our backs. I resent this and allow myself to grow angry about it, but I have no will to shout back or express my resentment in any other way. I find that if I give into emotion, or into panicked thoughts, like whether Matius escaped the plague, I grow weary enough to have to stop. Ahead of me, a figure in a hooded cloak sits huddled alongside the road, hugging their upper body as though in pain. It appears to me that he or she might have caught the plague. I resolve to stay clear of whoever it is.

When the huddled figure sees me pass, though, she calls out to me. A subdued voice quavers, “I have been struck by a rock or piece of wood thrown up by one of these fast-moving wagons. Please help me.”

As I turn to approach the huddled person, she looks up at me, crying in pain. I note that she is young, perhaps very young, and that she wears a white hood inside her outer cloak. She rubs her side below her left breast and asks for help in a foreign-accented German. I see a piece of wood as large as her leg laying in the road nearby.

I cannot demand that this stranger show me her underarms or her groin before I help her, but I see no visible buboes or other evidence of plague. I lift the hooded girl to her feet and struggle to help her move farther off the road and behind some bushes, so I can look at the wound. I pull the cloak and her inner garments, both ripped where the wood had struck the girl, aside. There is a grisly black and blue patch of skin, but no blood. I have nothing to help the pain, but the wounded girl seems to calm down as I run my fingers over the bruise, and then lay my palm on the black and blue mark.

The girl seems to take comfort from my touch.

In careful German, I ask how the girl should be called.

"My name is Sophia," she says. "I am a nun with the Catholic Church in Lübeck."

Sophia looks carefully at me. She searches my face for any reaction to her words, but I suppose she sees none, as she continues.

"Four days ago, my mother superior told me to take a few of my most precious belongings and some food, and walk as fast as I could to the Harz mountains. She said that a plague ship would arrive in Hamburg in five days. I became very frightened by her prophecy and left right away."

"I think that ship will arrive in Hamburg sometime today," I say, "unless the city leaders can persuade whoever is sailing the ship to stay away. I don't think that will stop the plague, but each day they are stopped from docking is another day the plague can't begin to move through the city. Can you walk?"

"Yes. But not as fast as I would like. Also, this parchment is heavy and unwieldy." Sophia opens her cloak to show me a pouch she has tied to the left side of her body. The pouch had probably stopped the wood from puncturing the skin over her ribs. In the pouch are some pages with writing on them.

"What is written on these pages?" I ask.

"These are words written by a church scribe in my monastery in Italy, in Florence. His name is Teodoro. He was a scribe to

Boccaccio in Ravenna. He copied some of the stories Boccaccio started to write about the plague. Do you know Boccaccio?"

I shake my head no.

"Well, two years ago, when the plague first came into the ports of Naples and Messina, there was a lot of panic all over Italy, among all the classes. Boccaccio wanted to capture the thoughts of some noblemen and women of that area as they tried to think how to escape."

I look at one of the scripts, but I cannot make any sense of it. It was written in Italian, first of all. "Can you read this to me?" I say.

"I will try to translate this," Sophia says, "but you may have to help me find the correct German word."

I get Sophia back on her feet and follow closely behind her as she hobbles forward. We go on in this way for another couple of miles, when the sun gets close to the horizon. "How do your ribs feel?" I ask.

"They hurt terribly, but I can walk with it," Sophia replies.

I can see the grimace on her face as the nun speaks. We are both tired, and we need to rest. We will need to stop before it gets too dark. We need to find a safe place to sleep. I also know that two young women on the road without any visible escorts are an invitation to trouble. I know myself to be strong enough and smart enough to discourage most fainthearted attackers, but I am not sure what I would do if I had to defend both of us against an attacker who is dedicated, smart, and hungry for whatever two women might have to offer.

I slow down to allow Sophia to take a slower, less painful pace. As we trudge along, I look around the shallow hills on either side of the road in the hopes of finding some shelter. Two mature ravens fly over them toward some trees to the right. One of the ravens seems to turn its head to look at me but then turns back to continue on its way with its mate.

At the sight of the ravens, I feel a twinge of longing for RabanHaven, on the shores of the Baltic.

I notice that the birds were flying toward a small copse of trees to the west that could provide not only small animals but also two traveling women shelter from the wind. We have been moving fairly fast on the road, and the crowd has thinned considerably as night closes in.

I keep walking to let a little more light fade from the sky. On an impulse I grab Sophia's arm and push her off the road, across a ditch, and up a small hill on the other side. We then crouch to walk furtively toward the group of trees where I hope we can find some shelter for the night. I hope that no one has seen us leave the road, but the crunching of nearby branches underfoot make it sound like someone has.

I whisper to Sophia to stop for a moment and sit quietly. As she does, I turn to observe the other figure so I could decide whether they might be a threat to us.

Before I can make such an assessment, the other gives away his position. He calls out to us in an accented German from about twenty paces away: "Hello. I don't mean you any harm. I only hope to find someone to travel with. I think we will do better sharing what we know, and whatever else we have as we try to get to the mountains. I have some food and a goatskin of water."

I whisper hoarsely, "Who are you?"

"I am Meshek. I take care of the accounts for one of the members of the Hansa who sells saddles and harness leathers in Hamburg. I hope the plague passes by without too much harm so I can go back to work there."

"What will you do if the plague lingers?" I ask.

"Then I think we are all in trouble," Meshek says. "But even that terrible eventuality might still be a good reason to share our resources. May I ask your names?"

I decide that Meshek seems trustworthy enough to get this bit of information, so I bob my head. “Yes. I am Agna, and my friend is Sophia. We are both strong and well-armed with knives and clubs. How do we make sure you are no threat to us?”

“What kind of assurance do you need?” asks Meshek. “The only real threat any of us has is the threat that one or all of us will catch the plague. I’d say that, if we have made it this far from Hamburg today, and we are each still in good health, then we are probably not a threat to each other from plague sickness. Now you say you are well armed and strong. I am only armed with food and water. So you have me at a disadvantage. Are you hungry?”

I walk cautiously toward him. “Yes. We have some food, but not much. Do you have enough to share?”

“All of our food will need to be carefully rationed,” he says, “but I am willing to put my whole stock together with yours so we can share equally. You say you have knives and clubs and the strength to use them. You should put yourselves at ease, then. Though I am a rather tall man, I have no weapons at all and have never been in a fight in my life. I keep books after all.”

Sophia speaks from her spot on the ground. “You sound like a good and careful man, Meshek. I think, Agna, that we can trust him.”

“Yes. All right,” I say. “But please walk ahead of us and a little distance away so we can get used to traveling with you. We are headed toward that small copse of trees up there. I want to get there before it is totally dark so we can see if it will work as shelter for the three of us.”

With our tall new traveling companion, Sophia and I make our way to the trees, arriving there in time to look around. The place seems suitable, so we move some branches around to give ourselves some comfort. I share some of my bread with the others, and Meshek passed his goatskin of water around.

Two ravens, one with a single white feather atop its right wing, come around to see if there might be a handout for them as well. I am certain that they're the same two ravens that guided me toward these trees, and I gratefully toss a small chunk of bread to each of them. Ravens are powerful symbols in the north. It would never do to fail to offer them sustenance.

"Your accent is strange to my ear, Meshek. Where are you from?" Sophia asks.

"I am Jewish," Meshek says, "which means that I am from many places. And you are Catholic. I can tell from your white coif, and what is left of the black in that part of your robe not covered in mud. What is your order?"

"I am Benedictine. My mother superior and I left our church in Ravenna, Italy, to escape the plague there, but now it has caught up to us again. By an arrangement with the bishop in Hamburg, we had been working at the Cathedral of Saint Mary. On hearing of the plague ships, my mother superior felt she must stay in Hamburg to help the sick there, but she told me that if I did not leave immediately for the Harz mountains she would make sure that I would be thrown out of the Benedictine order, and never allowed to return.

" There is a cathedral at the foothills to the mountains called the Walkenried Abbey. Mother Superior has given me a note to the Cistercian monastery there, to ask them to take me in."

Do you think they might take in a Jewish man?" asks Meshek.

The Cistercian Order, and my order, the Benedictines, are not allowed to turn travelers away. We must give them shelter and food, and we must bind their wounds until they are well enough to be on their way. You should have no fear of them."

"I am grateful to hear this," Meshek says, and I can hear the calm that enters his voice.

We soon turn in for the night, and weary from our travels, all fall into a deep sleep.

***

I wake hearing noises along the road. The sky along the eastern horizon is turning gray with approaching sunlight. I rouse my travel mates.

At Sophia's half-waking protests, I insist: "We should be on our way while the road is still mostly uncrowded. Here is a bit of bread to chew on."

Meshek passes the goatskin of water around and contributes a bite of rabbit meat to the common table. "We will need to find some food soon," he says. "I have a sling and a knife, and I am a pretty good shot with the sling. If you see any rabbits, or any abandoned animals of any kind that doesn't look sick, let me know. If they are nearby, I may be able to cripple or kill it with my sling."

"I think we need to look for a way toward the mountains that will be less crowded and less dangerous than this road," I say. "I fear that recent escapees from Hamburg may have been exposed to plague, and they may pass it to one of us. How are you feeling, Sophia?"

Sophia has her body turned away from us. Her head is bowed as she makes her morning prayers. She remains silent for a few moments, then turns toward us. "My side is still painful, but I think I can keep up with you through the day. Do you want to go now?"

"I think we need to start soon, especially if we are to find an alternative to walking on the road."

"I think we will make much better time on this road," Meshek says. "For one thing, we know where this road goes and we might find others, like ourselves, who have food and water and would prefer to travel in a group. If we go across the fields that I see here, we will spend a lot of time in mud and thick brush."

"I don't know if I could handle that," Sophia says. Her eyes are wide in fear and honesty.

Meshek hums in his throat. "Also, if we find a road, we won't know where it goes unless there are already refugees upon it. If there are refugees on any new road that we find, then we will have wasted the time we could have spent walking on the road we already know, and the progress we could have made toward higher elevations."

"You make compelling arguments," I say, "but I am still concerned that this road is the main road to the mountains, and it will become very crowded as the day goes on. I would have no doubt about looking for another route if it were just me and Sophia, but Meshek, your presence gives me some hope that threats against us will be less while you are here. But you have told us that you are unarmed. You have suggested that you are a coward when presented with the possibility of physical conflict, and yet, this morning you have told me that you are armed with a sling and a knife. So, please help me understand. Will you shrink from a fight and run away at the threat of violence, or will you stand to protect us?"

Meshek thinks for a moment, then rises. "I will answer your question, Agna, but we should get moving, and I propose that we stay on this road for now. As I said last night, the most important thing to me is traveling in a group. I am prepared to give whatever I have to any group that I believe I am compatible with, as I believe you both are.

"I do have more to contribute to such a group than what I let on last night. I am skilled with personal combat using knives and pugil sticks, or whatever heavy, blunt instrument is at hand."

Sophia is listening intently to this conversation with tears in her eyes. This morning, her native Italian dialect more deeply colors everything she wants to say. "I can't help in any of this. I left Hamburg with almost no food or water; only whatever scraps Mother Superior was able to give me as she brought me to the road before us now. I am completely dependent on you both to help me, especially now that I am in pain. Please stay with me; please help me."

Meshek looks toward Sophia for a moment, then back to me. He says nothing more, but we both rise to walk with him back to the road. Meshek stays to the side and a little ahead of us. Though he says nothing of the sort, he seems to believe that his presence will be more valuable to us if he is not seen in public as part of our party.

As we regain the road, the sun rises fully above the horizon. I expect the day to be hot. Meshek seems to have water, still, in his goatskin. I have a little as well, but we will certainly need to find water tomorrow. I believe that we will need another three days after this one to get to the edge of the woods that mark the foothills of the Harz mountains.

As we walk, Sophia tells us stories to keep our minds off of our aching feet. She tells me of Walkenried Abbey and its Cistercian monks and nuns. The Cistercians are associated with the Benedictines and share the values of hard work, particularly in agriculture, and particularly in their home communities. They run mines in the mountains, and they use the proceeds of their labors to help finance some of their commercial activities with the Hansa in Lübeck and Hamburg. It is rumored that they even used some of their funds to aid Pope Clement VI, who conducted his papacy in Avignon in southern France. Because of this, the church leaves them alone to work out their necessary arrangements with the Hansa.

The Cistercians have helped to finance some of the Crusades, and some of the Crusader knights financed by them had fought in the Battle of Leignitz. I flinch and avoid the impulse to tell this seemingly trustworthy sister of my own family's involvement in that long-ago fight. I do not know how she would take it or if she would turn on me.

"I am offering these insights to you," Sophia says, "so that you will see some value in my presence with you."

I grip her hand. "Please don't worry about that, Sophia. I am happy to help you on this journey that is so important for all of us. I know you would do the same for me, and probably even for Meshek. You must not think that either of us will try to find you someday after this is all over to lay claim against you for services rendered during this terrible plague. I can't even imagine such a thing in the ruins of society that will follow the end of this plague."

Meshek has been walking behind us, but now catches up. "I think we should stop for a bit to rest and eat something. Let's move off the road so we can have some privacy. I think I saw another rabbit. If I can catch it, I won't want to share with anyone but the two of you."

We three move to a small mound of earth in a cluster of small trees about fifty paces off the road. Meshek whispers that he will return short-ly. Sophie expresses her relief as she thumps onto the ground by smiling slightly in my direction through her pain and fatigue.

As I suspected, the road has grown more crowded as the day progresses. There are more walkers carrying their bundles. They look tired and downtrodden, especially when the wagons of the wealthier commercial classes pass by, kicking up rocks and the layers of dust covering the road, a dry, powdery dust drained of moisture by the hot sun.

Shortly, Meshek returns with a dead rabbit. "I brought this little fellow down with my sling. He didn't go willingly, but I'm sure he will be grateful to have given us needed sustenance on our journey."

"How good are you with that sling?" Sophia asks. "Are you like David taking on Goliath in the Valley of Elah?"

"Well, we Jewish people don't like to take on airs, so I will defer any comparisons with that David. However, I am pretty good with the sling, and I have had to take down the occasional hostile threat with it."

"I think you need to tell us more about who you are," I say as Meshek stows the rabbit in his pack, to skin and roast later tonight. "I believe you are a good man, but I also believe you have a troubled, or perhaps a troubling, past. Do you? Have such a past, I mean?"

His jaw tightens as he fastens his pack. "I have a past, as we all do. As a Jewish man who believes in his religion, though perhaps not in the rigorous practice of it, I have been many places and done many things to survive in a Christian world that is largely hostile to me. I have learned what I know of the sling from the story of King David, and I have learned what I need to know of combat without weapons from the Asians, who, unlike Christians, neither fear us nor seek to destroy us as a people.

"I could have made myself very much at home with the Mongols, if I had lived in those times, and if I had not found a good position with some businessmen in Hamburg who valued my skills and talent more than they feared my Jewishness. Actually, my final skill, with knives and other piercing instruments, I learned from these businessmen. Their specialty in Hamburg's commercial world is the cutting and stitching of heavy leathers used in saddles and harnesses. Having these leatherworking tools, and knowing how to keep them sharp, clean, and ready for use, is a matter of strong pride for me."

He frowns and pauses for a long moment. "You have many reasons to fear my skills in the violent arts," he says finally, "but, I think, many more reasons to view me as your protector, as that is my intent with you and with Sophia. I know Sophia is a member of a Christian order, though not one with violent intent toward the Jews."

"That is true," Sophia says. "We, the Benedictines and the Cistercians, are more interested in helping things grow into healthy maturity than we are in seeking revenge against ancient enemies as many Christian sects are."

Meshek nods in Sophia's direction. "Perhaps we should be moving along."

"Not yet," she says. "I want to say something. I do have something to help us in our journey besides a possible place to go to in the mountains. I've shown this to Agna, Meshek, but I want to show it to you as well."

He sits near her, and she begins her tale.

"When Mother Superior and I left Ravenna to escape the plague two years ago, we took some of our personal artwork as bargaining material for food or shelter or whatever we might need as we made our way overland to northern Germany. Our order in Ravenna had a very good reputation among the arts community there, and artists who had received favors from us would often repay us with their works. Some of those gifts went to the church, but some went to Mother Superior and I, because of our work in the fields for our order.

"Signor Alighieri writing his Divine Commedia in spoken Italian inspired many of the artists writing and painting more than a hundred years ago. Since the beginning of this century, the beautiful paintings of Dante's contemporary and friend, Giotto, have greatly influenced the artworks not only of visual artists around Florence, but even the writings of our good friend in Ravenna, Giovanni Boccaccio.

"As we were getting ready to depart Ravenna, rumors of plague were everywhere. Though we were making haste to leave, Boccaccio seemed to relish all the rumors of the plague's grisly effects, and the feeble efforts of the wealthy to not succumb to an illness of the poor. He began to write some stories, and he gave me a draft of one of those on paper. Even the paper itself was expensive, but nothing compared to the value of anything written by Boccaccio.

"I have this draft in my pouch, along with pieces of work from all of the beloved artists I have mentioned. I even have a small sketch done in ink on parchment by Giotto and given to his patron in those early days. His patron later came on hard times. Our order gave him food and a place to stay, and he repaid us with Giotto's beautiful sketch. My mother superior insisted that I take the sketch with me for safekeeping as we traveled north from Ravenna.

"When it was time for me to leave Hamburg, Mother Superior would not hear of me giving the sketch back to her. Tears of pain and longing came to her eyes, but she would not accept the sketch from me. She turned her back on me, and it was then that she threatened to have me thrown out of the order, something I could not bear.

"I still have the sketch with me today. I have been able to keep it with me for these two years, but I fear the loss of it now that the plague stalks us, as it stalks the desperate people who walk beside us along this terrible road.

"Everything in this bag is a treasure worth, altogether, an unimaginable value in the right time and place, but for now, I would almost trade all of it for safe passage to the door of Walkenried Abbey."

Sophia's head droops, now that she has exhausted herself of all she wanted to say.

I put my hand on her shoulder, then pull her toward my body in a warm hug. I think of how different I once saw us from the upper classes fleeing the plague in their horse-drawn carriages. Now I see that helping a stranger on the roadside has given all of us wealth and safe passage in the towns to come.

Meshek reaches out to touch each of us on the shoulder. “Thank you for trusting us enough to tell us this,” he says. “Now we must go.”

“There is one more thing,” Sophia says. She pulls a blue/violet stone out of the depths of her bag. It is no more than the size of my thumb, and the stone shines with an almost translucent beauty. The color changes as she turns the stone in the light. Neither Meshek nor I have ever seen anything like it.

“My mother superior gave me this stone and asked that I try to return it to the abbey in Ravenna,” she says. “Neither she nor I have any idea of its value but were told that it had been carried across the world and across time. Mother Superior told me the stone is meant to reside in God’s house. If anything happens to me, the stone is yours to do with as you please.”

With this solemn oath, we rise and rejoin the road south, away from the Black Death creeping along behind us.

On the third day of our journey, I once again rise first as the sun begins to turn the horizon slightly gray. This morning, though, the gray seems to have a reddish tinge, a sign of almost certain rain.

I shake the other two awake. “Get up. We need to make as much progress as we can before the rain gets too close to us. We will need to find a shelter from it, or we will need to make a shelter from whatever we can find in the woods.”

Fortunately, our increasing elevation as we reach the foothills has produced a few more trees along our route. The trees provide shelter and warmth, even the possibility of having some rabbit meat cooked over a fire. The idea makes my mouth water. The other two are still too groggy from waking up to care.

The rain comes down heavily for an hour or two in the morning. Then, the clouds clear, and the sun comes out to once again dry the roads. By afternoon, the winds kick up the dust from the roads, and it seems that each particle is destined to get into the eyes and nose of each traveler.

Soon we resolve to find a dry place out of the wind, so we can spend a day finding some meat and some firewood to cook it with. If we could find a stream coming down the mountains, we can even consider cleaning our grimy bodies. Once we find just such a place, we cannot help but feel overwhelmed with our good fortune in having survived this far and having each met and made friends with two good and supportive people.

Our good fortune continues after our day of rest, when Meshek hails a fellow driving a wagon pulled by two oxen. In the back of the wagon are two barrels of salted herring for delivery to the very abbey we are heading to.

On seeing the teamster's wagon and barrels of herring I cannot help but think of Kurt and the help he gave me on the muddy road from Lübeck to Hamburg. I still remember the two knights on their fine horses and how much fear they carried with them like a shield. I turn away from my friends to wipe away a tear that I might never see Kurt, or any of my family, again.

Meshek recognized the harness on the oxen as it was sold by the company he worked for. When he points this out to the teamster, the teamster invites the three of us to ride along with him for the remainder of our journey.

Some days later, we arrive at the door of the beautiful abbey. Sophia's letter is gladly accepted, as are the two barrels of salted herring. We are all invited to join the Cistercians at their evening meal.

Before entering the building, I look around the surrounding woods in the hopes I would see some kind of vision that would help me decide what to do next. Sophia is walking next to a monk of the abbey. She turns and looks back at me and smiles.

Meshek also looks at the surrounding woods, the sky, and the abbey grounds. As his eye catches mine, he, too, smiles.

END

Rebirth

Summer, 1475 CE. Orabella, a once homeless young woman now apprenticed to the Studio of Verrocchio, paints a fresco for the home of Lorenzo de Medici in the hills above the Tuscan city of Florence. She has the help of Leonardo da Vinci in the design of the work.

====

She is an orphan, a girl of the streets living in a poor section of Florence, near the industrial docks along the Arno. She does not know her parents, nor can she say with any certainty the last time she felt close to a grown man or woman, other than the young woman, a prostitute catering to the upper classes, who took her in when she was barely more than a toddler.

Their paths had crossed early one morning on Florence's back streets, as the taverns closed and before the shop stalls opened. The orphan had suddenly appeared beside the prostitute and taken her hand. The orphan looked up into the eyes of the prostitute who was, at first, not sure what to do with this small child. As she walked to her small but clean room after a night of entertaining the guests of her most important client, she was not at all sure she had the energy to deal with this.

After succumbing to the charm and pathos and need for decent clothes and a meal in the orphan's eyes that morning, the prostitute—she was called Magdalena—cared for the girl, teaching her the alphabet and basic reading and writing. She often took her around to see the sights, including the statues along the major piazzi and the churches in the wealthier neighborhoods, though, thinking it wasn't her place, she never gave the child a name.

Unfortunately, after a few years, the young prostitute grew ill and died in the orphan's young and caring arms.

After being turned out of the comfort of the prostitute's chambers, the orphan did not know where to turn. She had to develop intelligence, wit, and the physical and mental strength needed to survive. For a long time, her closest friends were other children of the streets. Most had the skills and toughness of the orphan, but none had her powerful desire to rise above their squalid lives. The child dreamed of rising, transformed, into the beautiful world Magdalena had exposed her to in her earliest years.

***

The girl does not have a clear memory of her former life until, one day, she and her young friends are walking along the streets toward one of the city's cleaner public markets. She is impressed by all of the fine clothes and refined speaking of the shoppers out on this sunny day. Though it has been a few years, a vision of Magdalena crosses in front of her eyes. She has to turn away from her mates, her eyes suddenly flooded with tears. Once she composes herself, the group goes back to the important business of lifting valuables from the purses and pockets of the wealthy.

In their frequent forays into the more well-to-do parts of the city, the knowledge the girl has gained from walking tours with her now-deceased mentor become a guide to the most crowded areas, where the pickings will be particularly fruitful for her tiny band of pickpockets and petty thieves.

After a few harrowing escapes across rooftops and through back alleys, with tradesmen, servants, and armed guards in pursuit, her young friends grow to respect all that she knows that is important to their independence and well-being. She knows the wealthy neighborhoods well, and she knows how to avoid the scrutiny of the many private security guards in them, who are constantly on the lookout for ragamuffins. The girl's friends soon begin to call her "Orabella," which means beautiful gold.

Other than the many sweet endearments from Magdalena, Orabella is the first name the girl has ever had that she can call her own. The young street tough who first gave her the name gave it with an attitude of friendliness that she did not understand. She had shown him no special favor. The tough—his name was Carlo—was overweight, dressed poorly, had pimples and a vulgar mouth.

Even so, Orabella could not ignore the fact that Carlo seemed to have the respect of the others. He was the one who collected the day's stolen goods and redistributed them to those who, in his judgment, had the greatest need for whichever item had been taken. Any leftovers were held by Carlo to be distributed on another day in response to a different and more urgent need.

Nobody argued about this for at least as long as Orabella had been with them.

Carlo protected her whenever any of the others tried to physically push her around or talk dirty to her or call her ugly names. When Carlo stepped into a situation involving Orabella, the others always stepped away.

***

Florentine artist and studio master Andrea del Verrocchio is working with a group of his apprentices on a commissioned work for Lorenzo de Medici.

The work, a large fresco to be installed on a great wall in one of the Medici estate houses in the hills north of Florence, is proving to be very troublesome, even at this early design stage.

As master of the workshop, Verrocchio, in the middle of his attempt to, once again, mediate the heated disagreements among his headstrong apprentices on how to proceed with the design work, is interrupted by a messenger from his client. The messenger, Adolfo Antonio, is observing the discussion from a corner of the workshop. He motions to the master to join him in a private conversation away from his apprentices.

"Greetings, my good friend," he says. "My master has become aware of the problems you are having with the design work on this project. He wishes to help in any way that he can."

"I believe our problems and disagreements are not out of the ordinary for a project of this importance," Verrocchio says, "but..." He pauses for a moment as he notices that the messenger has something specific in mind. "Please. I very much want to hear of any suggestions your master may wish to make to me, as well as any requests he may wish to make of me."

Adolfo responds. "My master has a young woman working as an assistant cook and maid in his household. In his mind, her real talents are greatly wasted chopping tomatoes, mashing eggplant, and scrubbing floors."

"I can't imagine your master even taking note of a household maid, unless she is a truly extraordinary person. What are these talents that can draw the attention of such a powerful and honorable man?" Verrocchio asks.

"Once she was cleaned up and given some clothes and basic manners, it turned out she is a very pretty girl, and very intelligent. She came to us an orphaned waif found on the streets by a friend who thought our family could help foster her innate talent and character. Thus, within the security of my master's household, our friend believes the orphan can become a more useful citizen."

"Why does your master think she will be any help to me?" Verrocchio asks. "If she is as pretty as you say, she will, for that reason alone, cause far more problems in the workshop than she will be able to resolve no matter how talented she is. I have never allowed women into these studios no matter what their skills are. Among these headstrong male apprentices, who are often more interested in each other than any woman can hope for, she will either distract them from their work or she will trigger the kind of male competition that can result in violence and the destruction of my property. Has your master given thought to any of these things?"

Verrocchio, realizing that he is being rude, does not wait for Adolfo to answer.

"In any case, I cannot see how her presence can help resolve our artistic problems in the way of completing this fresco to your master's satisfaction. How long has she been working in the de Medici household?"

"She has been with us for over a year," Adolfo replies. "She learns very quickly. She progressed rapidly to making desserts, even designing the decorations on cakes. Her creations in the kitchen have been prized by my master. He has said he cannot get enough of them.

"Even more importantly, she progressed into a supervisory position among the cooking staff without losing their support for her creative work. She had as much success with the household staff. She was even able to dissolve some tensions that built up over a period of months to everyone's satisfaction."

Verrocchio folds his arms. "You have given me reasons to further consider your master's suggestion, but I have one last question. Why would your master be willing to part with someone who is such an obvious asset to his household? Is there some problem you are not telling me about?"

"Whatever you may think of the motivations or intent of my master, you should know by now that he understands the careful development of natural talent, particularly if those talents emerge in members of his household. He recognizes the potential of this young girl, and he also recognizes that, with her abundant imagination, supervising a household will not hold her attention for long. He believes that the next step in her development is in your workshop. However, please understand that this is not a formal request. My master is well aware of the risks and difficulty of working with a headstrong young woman.

"He merely asks that you consider the idea. If you truly believe that her presence will so disturb your work on his commission that the work will suffer significant delays, he will not further pursue the idea with you."

"You and your master have given me much of interest to think about," Verrocchio says. "You have presented a strong case for my acceptance of this young girl even if she might make my life more difficult than it already is. By the way, what is her name?"

Adolfo replies. "She says a kindly old couple who took her in many years ago, now passed away, gave her the name Orabella; she does not know her surname. She has told me that she prefers to wait until she finds a surname she likes before taking one to keep for the rest of her life."

"A very wise choice. Does your master have a period of time in mind before I give my decision?"

"None that he has told me of but based on my overhearing your discussions with your apprentices, I should think it would be in your best interest to make a decision within the next two weeks."

Verrocchio nods. "I want to confer with my artistic associate Leonardo. If I decide to take this Orabella into my workshop, I will want to make sure that Leonardo is willing to share any risks of her presence, and will be at my side when I face the inevitable difficulties among my apprentices—even better, that he will be willing to give Orabella access to some of his own work to help in the development of whatever talent she might have to share with the painters' guild. As she grows in her own work, she will be able to give us a better idea of her real value to the workshop. Can you return in a week?"

"Yes," Adolfo replies. "My compliments on the wisdom of your decision in this matter, Andrea. Until then." He touches the brim of his hat, turns, and leaves the studio.

Verrocchio turns at the sound of a piece of plaster in one of the test frescos breaking and crashing to the stone floor. One of the apprentices has just expressed his criticism of a design offered by a rival by throwing a hammer at it. Verrocchio does not speculate on whether the hammer-thrower might have overheard his conversation with Adolfo. He merely wishes for a glass of wine and the end of his working day to soothe his growing headache.

***

Verrocchio gives a lot of thought to what he will say to Leonardo about Lorenzo's suggestion. To survive in his workshop, the girl will have to be extraordinarily capable as well as very assured of whatever skills she feels she possesses. When the inevitable artistic conflicts arise between her and the other apprentices, she will have to be fast on her feet, and she will have to win any contest, even the purely physical ones.

Verrocchio rubs the side of his rather large nose with his index finger as he tries to think how such a physical contest between a male and a female contender might end up. He has almost succeeded in talking himself out of the whole idea of bringing the girl into his workshop. It is too risky, and Leonardo will never agree to take on any part of the risk anyway.

***

Verrocchio, a very successful painter, sculptor, and goldsmith, is the master of the most successful art studio and workshop in Florence. He is a successful businessman and an astute judge of character. He attracts the best and most talented painters, sculptors, and craftsmen from all of northern Italy to his apprenticeship program. Leonardo, son of Piero, born twenty-two miles west of Florence in Vinci, is one of these. Verrocchio recognized his genius early, when he came to the workshop in 1466 as no more than a lad of fourteen.

Since the worst effects of the Black Death abated, it has become obvious to some that the old ways of regulating the social and physical passions of the community under the guidance of the Catholic church have collapsed down to their foundations. From the wreckage of the old ways of institutional expression new ways of individual self-expression have begun to flourish and, in individuals like Leonardo and the very young but very promising Michelangelo, to bloom with great beauty.

Verrocchio believes that Leonardo will, among his achievements in many fields, change visual art forever.

In 1472, the two men began their working relationship by collaborating on a painting for the Church of St. Salvi in Santa Verdiana. The painting, The Baptism of Christ, began as Verrocchio's commission, and he was to have been the primary painter, but Verrocchio was happier sculpting than painting. Among the other apprentices who would work on the painting, the very young Leonardo, still barely more than a teenager, was to have helped with the background and one of two figures in the middle ground—an angel holding the cloak of Jesus.

On seeing Leonardo's work on the painting, though, Verrocchio was pleased to discover that he had absorbed all of the stylistic brilliance of Giotto, the visual artist of a hundred years before. Leonardo was now extending Giotto's hard-won knowledge into new directions in Verrocchio's very own workshop. From this early example of Leonardo's genius forward, Verrocchio has been delighted beyond all measure.

Leonardo is a relatively tall man and was very imposing, headstrong, and rude to subordinates even when he was young. Many of the apprentices were angered by his abrupt manner of pointing out defects in their work. Fortunately for Verrocchio, they were able to mind their manners because they also recognized the precious gift of Leonardo's close scrutiny. A few carried grudges and sought revenge in ways that Verrocchio often found to be very disruptive and difficult to resolve if allowed to progress too far.

There is no question that Leonardo has many great insights and ideas. Following the introduction of Gutenberg's printing press in Germany several years before, Leonardo's use of now much cheaper and more commonly available paper for sketching in place of parchment is proving to be a godsend. These facts about the uses of paper are shown to be true, repeatedly, not only for Leonardo, but for the entire arts community in Florence.

After his meeting with Adolfo, Verrocchio is able to speak briefly with Leonardo a few times over the next several weeks. The two men are associated on several projects, and so there is a constant need for talk. When Verrocchio first mentions the young girl and the interests of their patron, Leonardo is dismissive, as though he does not want to be bothered.

Some days later, Leonardo himself brings up the subject. "The young woman you mentioned, Andrea. Have you decided to have her come here so that we can have a look at her?"

"I am surprised to hear this from you," Verrocchio says. "I thought you would reject the idea. The girl, Orabella, has impressed our patron with her skills both in the arts and in supervising the work of his household staff, but I have not met her, and I have not seen any of her sketches. Nor have I heard further from Lorenzo's messenger. I was prepared to send a message that we could not use her, but I delayed, waiting to hear your thoughts before I said anything at all. Shall I ask Adolfo to send her over with some of her artwork?"

"I doubt if fine clothes, artwork, and an introduction by a personal representative of the Medici will be a good introduction to our group here; or ours to her," Leonardo says. "Better, I think, if we introduce her as the daughter of an anonymous friend's family who would like to give her a chance to earn some money, and, maybe, learn something about drawing, painting, and sculpting. Introduce her as a kitchen helper and tell her she has a month to show her worth as a member of the staff."

Verrocchio realizes the political tact of this idea and flushes with gratitude. "I am amazed and pleased, Leonardo, that you have given this so much careful thought. I applaud your reasoning. In the spirit of getting her directly involved with our work here, I will propose to Adolfo that we send someone to escort her here by way of the marketplace so they can pick up some food for the household. That way we can get some idea of how she will work with our cooking staff, as well as how she will negotiate with some of the marketplace thieves who supply us."

"An important test. Who would you propose as an escort?" Leonardo asks.

"Vincenzo, the pigment grinder, has more experience in the market than anybody here. He is probably the best judge of the talents of a female buyer. He will probably want to pick up some pigments as well. A purchase of a small quantity of lapis lazuli will tell us something about her innate sense of the value of precious minerals.

"That will give us a lot of information to make our judgments on her suitability for the workshop, and for the apprentices' probable hostility toward her."

"Would you consider Amadeo as escort?" Leonardo asks. "The plaster maker? What value do you see in that?"

Listen, my very good friend. Since starting my own workshop, I have discovered that many of the commissions I receive include requests for a variety of small frescos. These are used as gifts, and for various household decorations.

"I know very little about frescos and have little interest in making them, but if your girl doesn't work well in your workshop, she may work very well in mine. She won't have to put up with the contending factions among the apprentices. If she is as fast a learner as you have suggested, then, within a month, she should have enough knowledge of plaster making, paint mixing, and fresco design to produce some of these minor artworks. In that role, she could be enormously valuable to me."

Verrocchio thinks a moment. "I will be more than happy to lay out this opportunity with Adolfo. I am sure he will be discrete and will carry it to Lorenzo de Medici with enthusiasm."

"Good. By the way, have Amadeo ask the girl—what did you say her name is?"

"Orabella."

"Ask her to pick up some sweets for the apprentices on her way over here. Maybe that will make them more receptive to her charms."

"Consider it done," Verrocchio says.

***

Amadeo the plaster maker is a little surprised to be assigned to escort a new kitchen helper to the workshop, even though she is to be escorted from the Medici household, perhaps the most prestigious address in all of Florence. He may have been insulted that he has to escort her through the food markets on a shopping trip, except that he is intensely loyal to Verrocchio and will do whatever he asks with all the talent and ability that he can bring to it.

"I'm told this is a very pretty and intelligent young woman," Andrea Verrocchio says to his long-time apprentice, "though still a bit coarse in her manners. Even so, when you present yourself at the door of the Medici, you will have a chance to actually see inside the home of the most powerful family in all of northern Italy, as well as the most important and generous of all the patrons who support our work."

"Yes, Andrea, but I am a little overwhelmed by it. You know that I am only one of the helpers here. I really have no training and no experience with this kind of work. Wouldn't Leonardo be a much better agent for you? I know he is a favorite of the Medici family."

"First of all, Amadeo, you are not just 'one of the helpers here.' I place great value in your experience and your work with plasters and paints. Without your work too many of our apprentices would be sitting around wondering what to do with all their pretty sketches and designs because they would have no materials with which to express themselves. In the situation with this young woman, you have more talent and interest to bring to this particular task than does Leonardo himself."

Amadeo waits, wide-eyed, so Verrocchio goes on. "As you know, Leonardo has learned a lot about fresco during the years he has been with us, but he has little interest in working with it. To be truthful, neither do I, but I do recognize that just about everyone in Florence with a piece of blank wall wants to own one of our plaster fresco panels. They will want the panel framed in wood with brightly colored images of the Virgin holding the baby Jesus, surrounded by cherubim and seraphim.

"A family wants a fresco that will last forever, that will hang on that wall in their home, covering up that blank space. I also know that we can create those panels, in quantity, in our workshop, under your supervision, but you will need more help.

"Now that Leonardo is trying to build business for his own studio, he and I agree that this girl may be able to learn enough fast enough to help us all produce more of these small frescos before our competitors recognize the value available in this new market."

"But if you and Leonardo are prepared to invest so much in a girl that you have never met, wouldn't you both want to escort the girl yourselves so as to get to know her better?" Amadeo asks. "She will have to be tested in some way, and quickly I assume, before her value can be judged at all."

Verrocchio pauses before answering. 'You would think so, wouldn't you? The problem is that we want this girl to start out as a kitchen helper and maid, and we will have to avoid giving her any impression of special favor. If the girl is to be of any value to either of us, she will have to rise on her own. I am assured by Lorenzo through his messenger Adolfo that she is fully capable of that if she has proper, and very subtle, encouragement and guidance.

"I would expect her to make herself available to you within a couple of months of starting her work here. When she does, you must be ready to assign your most disagreeable, dirty, and dangerous tasks. And you must be prepared to guide her through them so that she learns useful things without putting the health and safety of all of us, and the workshop itself, in great danger."

Verrocchio claps his hands together. "So, my friend, perhaps the most important part of your escort duty will be to impart the most interesting aspects of purchasing materials like limestone for plaster and other minerals for grinding into appropriate colors for fresco. If she seems worthy, let her work closely with you in the purchase of a quantity of lapis lazuli. I have a commission coming up that will require more than we presently have on hand.

"Perhaps you could let her actually select and purchase some mineral of lesser value. Your report of that, particularly on the quality, amount, and price of the actual purchase, will give us a greater understanding of Orabella's real talents than anything Leonardo or I can glean by talking with her."

"Didn't you also want me to take her to do some shopping for food supplies for the house?" Amadeo asks.

"Yes," Verrocchio responds. "Why don't you tell her that you have to buy some meat and vegetables for the evening meal, and that it will have to feed all of the household staff and apprentices—perhaps twenty-five people in all. Tell her that she is to decide what to purchase, and how to get the best price. If she does it to your satisfaction, give her the money to make the purchases. I will discuss all our plans with our cook, Giulietta. The cook will also be contributing her share of difficulties that will further test the mettle of this young girl. After all, she will have to survive the kitchen, before she will even get a glimpse of the workshop.

"I am relying on your judgment in all of this, Amadeo. You are a good man, and I believe you will do the right thing. Your work will make all of us proud. I now have other business to attend to. As soon as I have made arrangements with Adolfo for the date and time for you to appear at their door, I will alert you. Goodnight for now."

***

Several days later, Amadeo arrives at the Medici house. He straightens his shoulders, approaches the great wooden doors, and raps sharply. As the doors open, he feels his knees grow weak. He worries, briefly, that he might faint. The Medici are powerful beyond his wildest reckoning, and he wants so desperately to make a good impression. It is with a sigh of relief, then, that he greets the familiar figure of Adolfo, Lorenzo's personal messenger.

"Welcome, Amadeo," Adolfo says and waves him inside. "Wait here while I see if Orabella is ready. She does want to make a good impression on you."

Amadeo is a shy man. The idea that a pretty young girl would want to impress him is a little unsettling. The idea that she would accompany him through the very crowded and very public Florentine marketplace does not give him peace. Nevertheless, when Adolfo brings the girl in from another room, Amadeo stands taller and again straightens his shoulders the better to carry out his duties.

"Amadeo, may I present Orabella, a respected member of our household staff. We have very much enjoyed our time with her, and almost regret that she is leaving us. We take heart that the Verrocchio household is a great opportunity for her, and we wish her well."

Orabella is dressed in neat but plain clothes, suitable for a housemaid about to go outside on household business, but she carries herself with authority. She approaches Amadeo boldly with her hand outstretched, as if to shake hands. Amadeo has never heard of such a thing from a young woman. His words of greeting almost stumble incoherently out of his mouth, but her bold approach so overwhelms him that he puts out his hand to hers and shakes it.

"I am pleased to make your acquaintance, Orabella," he says. "Are you ready to apply your skills to the House of Verrocchio?"

Amadeo is pleased with the recovery of his dignity. Perhaps she didn't notice his initial discomfiture.

"And I am pleased to make yours," she says. "I am ready. May I call you Amadeo?"

"Yes. Of course."

Adolfo speaks up, obviously pleased by this initial encounter. "We will send a cart over to the workshop with Orabella's belongings later today. Is there anything more I can do for either of you?"

Amadeo bows, as does Orabella. Both offer a parting expression of warmth and gratitude, as virtually any Florentine would offer in response to an expression of favor by a senior member of the House of Medici.

Just as he is about to turn away, Amadeo thinks he sees the wife of Lorenzo, Clarice Orsini, looking on from a high balcony. Her interest is an obvious good omen.

He decides to start his tour with Orabella in one of the dockside shops along the Arno, where larger rocks from the distant quarries are broken down into smaller rocks suitable for specialized purposes, usually in the arts or other decorations. He can break these smaller rocks down further, crushing them for use as pigments when mixed in some kind of medium. These, he explains to Orabella, are also used in colored paints. Pigments made from minerals are very important in frescoes because minerals, unlike any plant or other biological material, will not degrade the bond between the dried plaster and the paint color. The degradation resulting from the use of plant and animal materials in paint will ruin a fresco over time.

Later that day, Amadeo and Orabella are in a butcher shop that sells pork and fowl, perusing the wares for that night's dinner. The shop is in the upper story of a building set into the Ponte Vecchio. Its windows look northeast along the river. Verrocchio is a prized client of his butchery, and so the shop's owner and principal butcher, Anselmo, greets Amadeo warmly.

"Greetings to you, Anselmo." Amadeo nods in return. "This is Orabella. She will be working with the kitchen staff in the Verrocchio household. I'm sure you will see more of her over the next several weeks. Please know that she will represent Signor Andrea as well as I do."

"I am very happy to meet you, Orabella." The butcher chooses this moment to wax poetic. "Bella, Bella Orabella. Amadeo, how do you rate the pleasure of the company of such a beautiful woman?"

Before Amadeo can answer, Orabella speaks on her own behalf. "I'm very happy to meet you, Anselmo. I can only hope that your meats are of the highest quality, as are your kind words, while preserving the most modest yet reasonable of prices."

Anselmo is dumbstruck by her words. In his world, no woman, and certainly no girl like this one, no matter how pretty, would talk this way on first meeting a gentleman shop owner. He looks in Amadeo's direction and winks. "Your young friend has a tongue, Anselmo. Should I always expect this kind of remark when she buys meat?"

Orabella immediately bows her head. "I am very sorry, sir, but I am new to this neighborhood. In my former neighborhood, I had to take a strident attitude when dealing with new shopkeepers, in order to establish the respect necessary to do my master's business. I can see that you are a different kind of person, a gentleman I can deal with." She turns to her escort. "I apologize to you as well, Amadeo. I did not mean to cause problems with one of the household's primary and most respected suppliers."

Amadeo cannot help smiling, though he does so discreetly, behind a hand, to avoid letting Anselmo see his enjoyment. "I'm sure we will all survive this encounter, Orabella. Let us find out if Anselmo can forgive your impertinence when you place an order for eight libbre of pork to be delivered this afternoon."

"Of course, Amadeo," the butcher says. "I will instruct my staff to wrap one of my best cuts of pork for delivery this afternoon, and I will instruct them to select and pluck only the plumpest of birds. Be assured that there are no hard feelings between myself and your young woman. Only a fool would feel anything but joy at the opportunity to work with such a pretty and forthright customer."

Orabella smiles broadly and shows all the charm she is capable of.

In her imagination she once again offers thanks to those in the Medici household who instructed her in manners, dress, and presentation. "May I examine the meat before you begin cutting? I would like to see how you go about your business."

Anselmo rolls his eyes toward the ceiling, then to Amadeo, then back to Orabella, before giving in. "Of course, you may. Please come with me."

Finally, their business completed, Amadeo and Orabella bid Anselmo a warm goodbye. They go to their next stop on the docks along the Arno west of the Ponte Vecchio. Amadeo needs to order quantities of various rocks for the colored pigments used in the studio's frescos.

Orabella's shopping trip with Amadeo is the most interesting thing to happen on that wonderful day of new experiences. She is out from under the benign but firm grip of the Medici household for the first time in many weeks. She almost feels as free as she was with her ragamuffin friends.

Amadeo believes that the most important part of his job escorting Orabella, in the eyes of Verrocchio, is the knowledge he is able to impart to her in his negotiations with the masons and rock smiths. They are the ones who cut and price the relatively small quantities of rock he needs for pigments. His job, and thus, Orabella's job, is to assure that the work of these rough men working with these rough materials meets all of the workshop's requirements.

All rock smiths are big men who are used to dealing with other big men in very physical ways. By the look on Amadeo's face when he looks at Orabella and winks it becomes clear to the girl that she will do just fine in her dealings with them. She smiles back at him. It is clear to her that the butcher shop owner, Anselmo, was a pushover compared with these tough men.

One of them approaches the pair. He is bold, his eyes traveling up and down Orabella's body. Though her long skirt hides her feet and ankles, her face and upper body and waist under her coat display her body in ways that can attract the scrutiny of certain men. Though she allows herself to blush a little, she does not turn away from the man's lingering gaze. Finally, he says to Amadeo: "Yeah?"

Amadeo does not appear to recognize the man, though he has done business with this rock yard several times over the years. They always have the stock he needs, and the price is always acceptable. "I am with the House of Verrocchio," he tells the new vendor. "This is my apprentice, Orabella. We need to purchase several varieties of rock to crush up for colored pigments."

The smith looks toward the ground and scratches his cheek. When he looks up, he seems confused. "What kind of rock do you need?"

"We can start with sixty libbre of limestone, then I need to see your stocks of ochers, siennas, umbers, terre verte, and any compounds of iron, or of manganese and iron that you have available. I need some greens, so if you have malachite, we will want to see that."

The smith says, "My name is Alfredo. Come this way. It may be muddy. The girl can wait here if she wants."

"She is my apprentice. She will come with us," Amadeo says, though Alfredo is already walking away.

"I'm pleased to meet you, Alfredo," Orabella says to his back. "I hope we will soon be doing business with you."

Alfredo looks back over his shoulder but says nothing.

The raw limestone is rough, but of sufficient quality to satisfy their purposes. Amadeo and Orabella pick through the other minerals and compare them for purity and quality. Once they have gathered what they need they ask that their purchases be delivered to the House of Verrocchio sometime within the coming week.

Alfredo quotes a price for the entire order including delivery. Amadeo and Orabella confer for a moment, and then Orabella offers fifteen per-cent less than Alfredo's quote. "I'll pay you a quarter of the total now, and the remainder on delivery," she says.

Alfredo's face grows red. He bends toward Orabella and speaks through clenched teeth: "I have offered you a good price. Do you think I am trying to cheat you?"

"No, Alfredo. Not at all. It is just that the rock samples you have shown us are of a lesser quality than I am used to. If you have samples of better quality, I will be happy to look at them, and will be happy to reconsider my offered price—if I believe they actually are of better quality, that is."

Alfredo looks toward Amadeo, but Amadeo has gone some distance away to look at different kinds of rock. "Wait here," Alfredo growls. He marches toward a small shed that might have been the rock yard owner's office. Shortly, he returns with another fellow who is not as big as Alfredo, but who does have a more commanding presence. "This is Matteo. He owns the yard and will be happy to discuss better quality samples and prices with you."

"It is my pleasure to meet you, Orabella, I think that is your name, yes?" Matteo asks.

"It is my pleasure as well, Matteo. Do you have some better samples I can look at?" Orabella asks.

"Yes, I do, but you will need to come into my office to see them."

Orabella hesitates at the invitation. She learned long ago that men who invite girls into private spaces sometimes mean to cause them harm. Her hesitation lasts only for a moment. Amadeo is some distance away, but Matteo does not appear to have any other motives toward her than selling some of his products. She follows as Matteo walks toward his office a short distance away. Once there he goes into a back room and brings out a tray of small samples of more carefully selected stones for her inspection.

Satisfied with their quality compared to the samples shown her by Alfredo, Orabella has no problem arriving at a price for each, and a time for delivery. Matteo also expresses no concern with her proposal for an initial payment today and a final payment to complete the transaction on delivery. At the end of it, Orabella offers her hand to Matteo to seal the deal.

Amadeo has given some part of the purse for today's purchases to Orabella so she can complete the purchase of the minerals to be used for pigments. She makes the initial payment to Matteo, then goes looking for Amadeo.

One more purchase remains, that of the very precious lapis lazuli, only available as an import by camel caravan from the Hindu Kush mountains. Shipments of anything, whether stone, carpets, or rare spices, from Afghanistan take several months over deserts and mountains from the east to Florence. For studios like Verrocchio's lapis is very important in certain works by very desirable and very influential patrons. Lapis lazuli is ground into ultramarine pigment to be used in painting the ceremonial robes of kings, emperors, and the Virgin Mary.

Matteo does not have the security necessary to protect a stock of lapis. He refers Orabella to a jeweler near the Medici bank who might be able to provide her with the precious material.

Amadeo has been walking among some of the piles of materials while Orabella conducts the business of the Verrocchio household. When she catches up to him, he has paused to feed some bread to a raven perched nearby. The raven puts a foot on the piece of bread and tears off a more edible chunk with his beak, then crouches and looks up from his eating to note her arrival. She notes a white feather on the upper right wing and asks Amadeo how long the raven has been with him.

"For several minutes now. The raven is a very intelligent animal, and this one seems to want to talk after each morsel of bread."

She tells Amadeo of her purchases, and her experiences with Matteo. "How did you learn to judge the quality and price of these lesser stones?" he asks.

"In the House of Medici, there are always people available who can examine and evaluate materials brought to the household for sale. Because they believe me to be nothing more than a curious girl, who might also have the favor of Lorenzo, these people would not hesitate to answer my innocent questions about their testing and evaluation methods.

"One of the Medici people knowledgeable about the price of lapis today says that the twelfth part of a libbra will have a current value of about eight florins, but you must check the daily price on the florins to be used for the transaction before agreeing on a price. As I am sure you know, the real problem in determining value in any sample of raw precious minerals is the inclusions that take away value, and the purity of what remains after the inclusions have been removed."

"We have dealt with the same dealer in lapis and other precious minerals for many years," Amadeo tells her. "I am sure he is trustworthy, but if you are able to test a sample of lapis for purity in a way that our dealer will agree with, we might be able to get a better price. That would please our master."

"I have never done such a test," she says. "If the need is urgent, we will have to buy the lapis today at the dealer's price. Once we are back at the workshop, I will try to find out more about how to do the testing to see if we are getting good value."

"Our jeweler's shop is near the studio," Amadeo says. "It is on our way back."

As they head to the jeweler's, a young man steps out of a doorway, onto the street, and almost bumps into Orabella. They both excuse themselves then continue on their way, but Orabella stops suddenly. She turns to look after the young man. Even from the back his purposeful walk is very familiar to her, but not his new and well-fitting clothes. "Pardon me, sir," she hails the young man. "Do I know you?"

He turns and looks toward her. "No, miss. I am sure not, and that is my great misfortune." He looks toward Amadeo then touches his forefinger to the brim of his hat. He gives Orabella a warm wink and broad smile then turns to continue on his way.

Orabella is momentarily puzzled. "I'm sorry, Amadeo. I thought I knew that man, but he is obviously a stranger to me. It is getting dark. We should keep going."

But the man's familiarity nags at her as they walk. Suddenly it occurs to her that the young man is taller, his voice deeper, than the last time they saw each other. He has lost some flesh and outgrown his pimply face, but she knows that it is surely Carlo. With that realization a new question nags at her: Where did he get the money to dress so well?

"Are you all right, Orabella?" Amadeo asks, unaware of any of this.

"You've suddenly grown pensive."

"I am fine, thank you. I think the long day of work and meeting so many new people have made me tired. I'll be better when I have some-thing to eat."

As they walked back to the studio, Orabella's mind wandered back to the early morning when Amadeo had arrived to take her from the house of Medici. Lorenzo's wife, Catherine Orsini, had appeared on her balcony briefly to offer a friendly but discrete wave to Orabella. The previous day Orabella had left a thumb-sized almost iridescent blue/violet stone on Catherine's dressing table. She had composed a note to Catherine expressing her humble thanks and her hoped for acceptance of this modest symbol of her gratitude for the fine treatment she had received from the Medici family from her very first day in the house.

Orabella had debated whether or not to tell Catherine where the stone had come from; Orabella's early mentor, Magdalena, had received the stone from one of her regular visitors from the City of London. Finally, she had decided not to tell anyone where she had acquired the beautiful stone.

***

The two shoppers complete their list and are able to return to the House of Verrocchio before nightfall. Amadeo introduces Orabella to the main cook, Giulietta, and to the master of the household staff, then takes his leave. Giulietta gives Orabella a clean but worn dress and apron and shows her to her small bed in a corner of the large pantry. Once changed into the work clothes, Orabella is to help with cleaning and cutting the vegetables, and with cooking the evening meal.

Giulietta is quick to find fault with her work, but Orabella is not surprised by her treatment as a common kitchen helper. Though Adolfo had not told her what she would be doing in her first days and weeks in Verrocchio's household, he did say that she should do every assigned task with the same selfless enthusiasm she had when she was first brought into the House of the Medici almost two years ago.

"If you can do that with good cheer," Adolfo had said earlier that day, as he walked with her to the entryway where Amadeo waited, "I have no doubt that you will rise quickly to more important work in the House of Verrocchio."

Minding Adolfo's advice, Orabella approaches her assigned kitchen tasks with great energy. Even on that first day, she can see some changes in the flow of kitchen work that she might suggest to the cook. She resolves to wait until the meal has been served, the diners satisfied, and the dishes cleaned to do so.

Or maybe, she thinks as a frustrated Giulietta clatters used plates into the water basin, I will wait for a few days before saying anything. She is sure the household staff will test her in some way, possibly many ways, and she will need to survive all their tests if she is to gain their respect. Until then, she decides it is better if she does not make any suggestions. She allows herself to become encouraged and hopeful about this testing period when she notices that Giulietta, in the midst of expressing a highly critical comment about the poor result of a particular task assigned to her, cannot quite suppress a smile.

Orabella works hard with the staff. They may have imposed tests on her, but the tests do not take away from her general cheerfulness. After several weeks both Giulietta and the staff supervisor look forward to working with her and seek her suggestions for how they might get more of their work done in less time. As good reports of her efforts are laid before Verocchio, he begins to think it is time to bring his newest apprentice into the workshop.

A few days later he visits Amadeo's studio. "I am told that our new girl is working out very well with the household staff. Have you heard any-thing to the contrary?"

"No, master," Amadeo says. "She is everything Adolfo promised. If you want to bring her into the workshop, I have no objection. When the time is right for her reassignment, I think it will be best to have her start by cleaning up the fresco studio and organizing the paints and materials in there. That will bring her into contact with the apprentices. I will give her plenty of room to work out any problems with the apprentices in the way she thinks best. I am confident that she will do well.

"When it is time," he continues, "I will choose a well-qualified apprentice to work with me on the new fresco commission. Orabella will assist us. If she works as well with him as she has done with the household staff, she will have my highest recommendation and my full support."

"Do you feel confident that you can pass your skills in fresco along to Orabella sufficiently that she can finish a complex work on her own should I, for example, ask you to take another project that might not involve frescos?"

Amadeo paused, curious to know what his master might have in mind. "I have no doubt that I can pass my skills along to her," he said, finally.

"Good. Thank you, Amadeo. You are a good man. I will discuss all of this with Leonardo. I am sure he will agree to our plan." With a wave toward Amadeo, Verrocchio returns to his private studio elsewhere in the workshop.

A few weeks later Leonardo visits the studio. "I am very pleased to hear that the young girl is working out so well with your household staff," Leonardo says to Verrocchio. "If she is able to deal with Amadeo's randy apprentices, and actually starts working on a client fresco, I will want to see her work in progress. Make sure that Amadeo gives her significant work in the preparation and mixing of color pigments, and in the preparation of a cartoon that can be shown to the client. I will be interested in seeing the quality of her work in those most difficult of fresco problems, and I will criticize any work she does accordingly."

"I am very confident, Leonardo, that she will pass any test of her artistry that you may want to give her, though she has not claimed to be an artist," Verrocchio says. "We may find that she does have that kind of talent, or we may find that she does not. In either case the real question for our Orabella is whether she can use the resources of the studio to produce frescos that satisfy the customers who are willing to buy them from either of our studios."

With a wave of his hand Leonardo left the building.

***

A week later, Amadeo comes into the kitchen, begs Giulietta's forgiveness for taking Orabella away from her chores, and asks Orabella to walk with him to the fresco studio. The cook and her helpers all wish the girl well, hug her, and ask her to come back to the kitchen whenever she wishes to visit.

Before leaving with Amadeo, Orabella goes to her space in the pantry to change into the rough cloth pants, smock, and hair cover that he has given her to wear in the studio. For now, she will return to this space when her work is done for the day. Eventually, if things work out, she hopes she will have a more private space closer to the studio.

Amadeo is clearly embarrassed as he talks to Orabella about the general teasing and harassment he expects she will suffer from one or more of his five apprentices. "They are quite full of themselves," he says, "and their occasionally rude and offensive behavior is quite beyond any criticism in their eyes. But I want you to criticize them to their faces when they do not do their work. The only authority you will have is your femininity and your may gain you some tolerance that you would not have if you were another male. Do you understand me?"

The blunt talk of sex and sexual temptation causes Orabella's mind to drift back to her chance meeting with Carlo a few weeks ago. She wonders what happened in his life to turn him from a chubby, pimply child into the self-assured young man with his warm and inviting wink.

Her early memories about her mentor Magdalena are usually vague. She remembers the many wonderful places they saw together, but not much of their time spent sitting and talking between lessons nor the times she spent waiting for the prostitute's return at odd hours of the day or night.

Orabella was certainly not ignorant of the grunting and sometimes naked coupling between men and women. It was not uncommon on the streets late at night, but it had not meant much to her until she saw Carlo again.

Her thoughts about Carlo's earlier protections, if left unchecked, often lead to a feeling of warmth in her torso, breasts, and deep in the pit of her stomach. These feelings give her a kind of soft pleasure that takes the edges off the uncertain things in her life that cause her to fear the future.

That night, after Amadeo's warnings, her mind wanders as if unchained, carried forward by her imagination. Her hands also wander, first, to her neck, where she lightly touches her throat, then her collarbone, before traveling down the slope of her breast to rub and squeeze her hardening nipple. Her hand ghosts down her body and around her hip to grab hold of her left buttock as though Carlo were holding and squeezing her there.

The other hand moves down her belly to her sex as she spreads her legs and begins to move her fingers and hips in a beautiful synchrony. As she continues these rhythms her legs spread further as though someone, as though Carlo, is easing them apart. The pleasure of the movements through her lower body grow in intensity. A low moan escapes her lips, but she cannot stop and does not want to stop the quickening motions of her hands and fingers and legs. Suddenly in an intense burst of pleasure her body and legs tighten around her right hand, her fingers still moving in time with her bucking hips. Her left hand moves away from her buttock to grip the bed's edge until the release seems to be complete. Eventually her hands fall away, and she lies back to catch her breath.

Orabella has never allowed these feelings to mature enough to cause this physical response. She wants to have them again, but the next time, she allows herself to imagine, she wants Carlo's hands in place of her own.

But how will that be possible? she wonders. She has only seen him briefly. She has no idea if he lives in the area, or if he was merely visiting. She clearly remembers the doorway and the building that Carlo emerged from when he almost knocked her down. She resolves to find out more about that place. If necessary, she will knock on the door and inquire if a young man lives there. If necessary, she will tell a small lie to the effect that she had seen him drop something while he was shopping, and the shopkeeper had directed her to this door.

***

For her part, Orabella takes extra care to button her studio clothes and wear them loosely, and to bind her breasts so as not to give any ideas to the apprentices.

She dealt with harassment before she was taken from the streets by the Medici, and she learned how to deal with most of it. When confronted with the possibility of violence, running quickly away usually took her out of harm's way. When this was not enough, she had no hesitation about using whatever stones, bricks, or heavy pieces of wood were at hand.

Amadeo tells her to report any threats made by any of the apprentices, but she knows that such threats can never be reported unless her life is in danger. Any such complaint that got back to the apprentices would mean the end of her gaining their trust and confidence. Even their tolerance for her continued presence in the studio could be at risk.

Giulietta has become a close friend and confidante, though, and she has an intimate knowledge of the internal politics among the staff and the apprentices. She can see that Orabella will need help in dealing with the hostile feelings between some of them.

Antonio Sforza has recently been brought into the House of Verrocchio as a favor to Lorenzo de Medici. Giulietta knows him to be serious trouble, and she knows that Orabella can get caught up in his manipulations if she is not made aware of what a bad person he is. Giulietta believes that Orabella can be an ally in easing Sforza out of the house sooner rather than later. That way the damage he might cause can be held to a minimum.

The apprentices' threat to Orabella comes within a few weeks and is as quickly resolved. It turns out that one of the apprentices in the fresco studio, Poldi, a helper in making plaster who has more looks and muscle than intelligence, took a bet offered by Antonio. Poldi, who often flaunts his good looks, said that he could persuade Orabella that she should have sex with him, and she would agree. Antonio bet that she would not do it, knowing that his challenge to Poldi's manhood would give Poldi added strength in actually convincing her to do it with him. Antonio promised to assist in the enterprise by persuading the other apprentices to leave the studio.

When the day arrives, Poldi tells Orabella that he needs her to help fire the kiln in order to render a batch of limestone into quicklime. She is always eager to help with these kinds of tasks so that she can gain as much experience in as many aspects of the workshop as possible.

"I have to go back into the studio for some tools," Poldi says. "Then I'll go to get some wood to get the fire going. Use the wheelbarrow to start bringing the limestone over there and put it next to the kiln."

He goes into the studio and catches Antonio's eye. In a few moments, the studio is empty. Poldi returns to the wood pile, takes his shirt off, picks up an ax, and begins splitting the wood into pieces that will burn with a fierce flame. "Orabella!" he calls. "Come help me split the wood when you have finished moving the limestone."

She does not hear what he says, but she has heard him say something. She goes to the back of the woodshed where he is working. Since he often takes his shirt off when working with the kiln, she does not at first think further about it.

"This wood is still green," he says. "I need some help separating the ax from it, and some help getting the split wood out of my way. Can you do that?"

"Yes, of course. Stand out of the way, and I'll clear the wood behind you."

As she walks behind him, he turns and puts his hand on her shoulder. "You know I have had my eye on you, Orabella, since you came to work in the studio. I think under those loose clothes you are probably..."

That is as far as Poldi gets in making the proposal that he was sure would win him the bet with Antonio.

Orabella's mind goes blank. In the blink of an eye, she turns toward Poldi, and kicks him, hard, in the testicles with the solid toe of her shoe. Poldi screams and falls to the ground rolling, crying, and screaming. He pulls his knees to his chest while holding onto his painful private parts. She kneels beside him. She demands that he get up and stop blubbering.

Poldi is barely able to sit up. His crying turns into a whimpering moan. He continues to hold his testicles, and he cannot sit comfortably because he is in such pain. Tears are streaming down his face.

"You should consider yourself lucky, Poldi," Orabella says. "I could easily have put out one of your eyes. You made a mistake, but I will forgive you for the insult if you will promise me this: do not ever talk about this to anyone, and especially do not talk to Antonio. Do not try to claim that you won the bet. If he asks you, tell him he will need to talk to me directly to find out what happened. I will dispose of Antonio in my own way. If I ever hear that you have discussed this misbehavior of yours with anybody, the swelling in your balls might never go away, and you might never again find work in the city of Florence."

She begins to turn away but whips back to him in a way that causes Pol-di to flinch. "There is one more thing. If any of the other apprentices ever approaches me with the same idea that you had a few moments ago, then you must walk over that person and hit him, hard, right in the face. Then walk away and go back to your work. Can you do that for me, Poldi?"

Poldi—still crouched over, moaning and sniffling—nods yes.

"I mean you no harm if you behave yourself, but if there are more of these challenges waiting for me anywhere in the workshop, I will need your help. Do we have a deal? If so, put out your hand." Poldi does so. Orabella shakes his outstretched, trembling hand. "All right. It is time to get up when you are ready. We have to finish preparing the limestone."

Amadeo had come into the studio to find out why all the apprentices were away. As he looks around, Poldi and Orabella walk into the studio from the Kiln. By the look on Poldi's red, teary face, and what Orabella hopes is her own calm demeanor, Amadeo knows exactly what happened. He is glad the apprentices got their sexual challenge out of their system, and that nobody was badly hurt. Poldi will recover.

"Hopefully," Amadeo mutters, sotto voce, "we can now focus on the work."

***

Leonardo and Verrocchio are very pleased by the reports they have received about Orabella's confrontation with Poldi. When they think of the expression on Poldi's face when he first walked into the studio to face Amadeo, both break into laughter.

"I believe Orabella has earned the right to supervise the creation of a fresco under the watchful eye of a paying client," Leonardo says, wiping a tear from his eye.

Verrocchio agrees. "She still needs some training in design. Amadeo has not yet had an opportunity to see if she can work with the client to come up with an acceptable one. Once she has developed a design for the client's inspection, I think it will be important that you come in to inspect her work."

Leonardo nods. “Stay in touch with Amadeo as she prepares the plaster for the application of the design. I want to know if she has any particular problems developing any part of the fresco.”

A few days later Amadeo gives Orabella the authority to supervise the apprentices in the creation of the new fresco. When he leaves the room, she asks them to gather round to discuss the course of the work. She first asks Poldi about the status of the current batch of quicklime, rendered shortly after the incident behind the woodshed. Antonio Sforza has been banished from the workshop, and Poldi has learned to be respectful and deferential around the female apprentice. She, in turn, has become courteous toward him in these kinds of meetings, and in her supervision of him.

“The quicklime will be ready for mixing whenever you are ready, miss,” Poldi says. “If we need more, Fons will help me fire the kiln. We have plenty of limestone.”

“Good. Thank you, Poldi. You are a good helper to me.”

Orabella is already beginning to feel guilty about the way she dominated Poldi in the woodshed. She wonders if she perhaps misunderstood his intent. She knows that Sforza, the lazy troublemaker, had put him up to it, but she is beginning to doubt that Poldi would have gone through with it. She directs her next comment to Vincenzo, the mixer of pigments: “Are there any problems grinding any of the rocks we need for the pigments, Vincenzo?”

“No, miss. As soon as Fedele and I have the cartoon with the design and colors, we will be better able to judge the suitability and quantities of the materials we have on hand. Pardon me, miss, but who will be doing the design?”

“I am told by Amadeo that Leonardo, though he does not have a particular interest in working with fresco, has taken an interest in this one.” As Orabella says this, the other apprentices sit up. “When the time is right,

I am told, he will work with us on a design that our customer will find acceptable. In the meantime, I will need to work with Elia on the creation of the cartoons that express the artist's design. Elia will work directly with Leonardo as he always has, but Amadeo has asked that I learn something of the creation of them in anticipation of future commissions."

Vincenzo speaks up again. "When he last spoke to us, Leonardo expressed some interest in exploring the use of linseed oil as a medium for the pigments rather than egg whites. He believes the oil—because it takes much longer to dry, weeks instead of minutes—will give him much more flexibility in composition. The problem is that we have been using egg tempura on parchment for at least a thousand years. We are very experienced with it, and I am a little concerned about Leonardo's wish that we break from these ancient traditions."

"Thank you for expressing your concern, Vincenzo," she says. "Have you done any work with linseed oil as a color medium?"

"Yes, miss. I have mixed two batches of pigment using the oil because Leonardo requested it. If he is now asking to use oil for this commission, I guess he was satisfied with the earlier results."

"Has either Amadeo or our master expressed an opinion on it?"

"No, miss. My understanding from both those gentlemen is that they want to give Leonardo as much freedom as possible in experimenting with new techniques, materials, and methods. That is, of course, unless the client specifically forbids experimentation beyond traditional practice. To my best knowledge, that is not the case here."

"Do you know of any particular problem with the use of egg tempura that suggests we should fully embrace the use of linseed oil?"

"Some have suggested that frescos that have relied on any compounds using organic matter like eggs or plant materials lose their color and tend to grow mold in the part of the fresco where they are used. I have not observed these problems myself, but they do give me pause in advocating their use. Linseed oil is also made up of plant material, so I'm not sure what we will gain by using it in fresco. Use of it will, of course, be very agreeable to Leonardo in that it will give him more time to detail his design."

Orabella is thoughtful for a few moments. "I don't want to make a decision on this, but I do want to keep the discussion going at least until we can hear more from Amadeo. I will ask Amadeo to confer further with the master, and with Leonardo. Hopefully, I will have a better answer within a few days. For now, I would like to get to work. Vitale, when can you have the frame done for our new fresco?"

"The frame can be ready later today, miss."

"Good. Poldi let's meet early tomorrow morning to look at Vitale's work. If the frame is ready, I want to lay the first binder coat of plaster so it can begin to dry."

***

When Leonardo visits the studio, he comes in good spirits and leaves in even better ones. Amadeo tells Orabella that Leonardo is so pleased with her work on the cartoons that he offered to help with the actual design work for the fresco. When the client hears of Leonardo da Vinci's interest in the work, he offers to double the commission.

Verrocchio and Leonardo are so pleased with Amadeo's reports of Orabella's work that she is almost certain that she has a guaranteed spot in either man's workshop. Both men feel the need to discuss these very positive feelings about their experience with Orabella to Lorenzo's representative in his dealings with the Verrocchio studio. They invite Adolfo Antonio to join the two of them for a quiet dinner at the studio later that day. They tell Antonio that Orabella will also join them for dessert after the dinner.

Between bites of lamb Adolfo tells the two men that Lorenzo already knows of the progress that Orabella has made under the guidance of Verrocchio and Leonardo in working with the fresco process. "As a matter of fact, "Adolfo notes, "Lorenzo has given me this interesting blue/violet stone which was presented to him a few months ago for one of his past good works. None of Lorenzo's appraisers have ever seen such a stone and, so, they have no idea of its' value. He asks, Andrea, if you can find somebody who is qualified and willing to appraise the stone. If you can find such a person, he will be willing to share in any value realized with you.

"This is, in part, an acknowledgement by Lorenza of his gratitude for the most excellent work you and Leonardo have done with Orabella," Adolfo said as he swirled a bit of bread in some lamb gravy on his plate and washed it down with a mouthful of red wine.

Verrocchio expressed his gratitude for Lorenzo's gift with the deepest possible humility, though, in truth, he had no idea where he could find a suitable appraiser.

As he had promised Adolfo Verrocchio asked Amadeo – who had been acting as sommelier at the dinner – to summon Orabella to join them for desserts. The three diners heaped such praise on her for her achievements that she had to excuse herself to wipe away her tears of gratitude on receiving such recognition from such distinguished Gentlemen of Florence. On her return to the table she sat down to a very fine cobbler drenched in sweet sauce sprinkled with pine nuts.

Orabella knows, now, that she must make amends to Poldi. She has already told Amadeo of her regret at the way she treated him like a common street harasser and the way the others treated him after the incident. She will apologize and make sure he can restore his good reputation among the other apprentices.

But, first, she will resolve the problem of finding Carlo. The nights since she saw him have been restless, relieved only by her own intense imaginings. She worries that she needs some resolution of these feelings if she is to have any lasting value to the House of Verrocchio. As to the nature of that resolution? She hopes it will include much more than sex, though she certainly hopes sex will be a major part.

The Blue/violet stone was not further mentioned at the dinner and Verrocchio was left to ponder the disposition of it. A few days later a Scottish gentlemen who had recently become wealthy in the wool trade and, following that, a regular buyer of art works produced by the studio engaged Verrocchio in a discussion of gemstones. Andrea saw his opportunity to get the stone into the hands of an appraiser in London who would, no doubt, have knowledge of this stone and of the value that it might fetch. An arrangement was agreed to; the Scottish gentleman took the stone but said he could make no guarantees about when he might be able to find an appraiser and submit a report.

Andrea, though he could certainly appreciate the beauty of the stone had no way to determine its's value or whether it had any value at all. Such mysteries do not occupy much of his time and attention, so he soon forgot about the stone.

***

On her next shopping trip Orabella is gone longer than usual, and Amadeo begins to worry about her. When she returns, he asks if she ran into any difficulties.

"No, Amadeo," she says. "I appreciate your concern, but I needed to spend some time by myself. So much has been happening to me lately that I needed some time to think.'

"Is there anything I can do?"

She gives a weak smile. "No. Not now but thank you."

"Well, if you feel concerns about things beyond the work, I can at least offer you an ear. In fact, I would be happy to. For now, though, I will let you get back to business. I will see you later at dinner."

As Amadeo takes his leave, Orabella cannot believe her luck in joining the House of Verrocchio.

What she does not tell him is that her shopping trip had grown longer because she hoped to learn more about Carlo's whereabouts. She is not ready to discuss anything about Carlo, or anything about her life before the Medici took her in. She went into the neighborhood where she had last seen Carlo but had no luck in finding him. She went up to the door he had come out of, paused for a moment, then boldly knocked.

The door opened slightly. An old man, skinny, wattled around the jaw and bald, looked out.

"What?" he asked bluntly.

"I'm looking for a young man who might have lost something in the market a couple of weeks ago. When I asked the shopkeeper about it, he suggested I come here, that you might know something about the man. Do you know of such a man?"

The old man opened the door wider and looked boldly over all of her body, from the top of her hat to the tips of her shoes. "Do you have a name for this guy?"

Orabella hesitated. She became suspicious that there might be more to Carlo's situation than might have been apparent when she bumped into him. She decided to find out more before she allowed her name and the name of Verrocchio to be associated with Carlo. "No. I don't," she said.

"You are a good-looking woman searching a run-down neighborhood for a man whose name you don't know. I think you had better spend your time on a different project that doesn't involve this address. Go away." He slammed the door.

Orabella sets her parcels on the counter, where Giulietta will find them later. She decided on the stoop to take the old man's advice, but she knows, deep within herself, that she will be back there soon.

END

The New World

Early spring, 1905 CE. Eirian Ross, a woman retiring from her post at Cavendish Laboratory, looks back over her bright career, which began when she was trained, surreptitiously, in advanced mathematics at the University of Edinburgh, after working with scientist James Clerk Maxwell at his rural estate in nearby Glenlair. Together, they will test and document the partial differential equations that Maxwell has developed to define electromagnetism. Eirian also reflects on her time studying the writings of Ada Lovelace and attending offbeat lectures by Charles Dodgson.

As she packs up her office to retire from the prestigious Cavendish Laboratory at the University of Cambridge, Eirian Ross reflects on her career as a woman in mathematics and science, especially in conjunction with scientist James Clerk Maxwell.

Maxwell had mastered the basics of science and mathematics from his earliest days at school. In 1865—well along in his career as a scientist, mathematician, and builder of lab equipment to support his experiments—he sought the help of farm girl Eirian Ross, his neighbor and now his protégé, in preparing and checking his experimental notes and papers on electromagnetism for publication.

Despite his deep investment in science Maxwell was a generous and caring man. His work with the Episcopal Church of Scotland, alongside his wife, Katherine, helped him to keep close contact with the world outside science, and with the human and personal needs of that world. Eirian had not been raised in a particularly religious household, but she was open to all of the ideas that gave persistence to the dogmas of the church, especially when her mentor James Clerk Maxwell articulated them for her benefit.

Unfortunately, in these, days, young women were not allowed to attend university as regular students. However, for Eirian and her dear friend Bethan, both eager for education beyond reading, writing, arithmetic, and homemaking, Maxwell promised his full support for their interests in learning at the University of Edinburgh, even though that school had never allowed female students. Maxwell assured both their parents that he would make arrangements for tutoring when necessary. He would supply them with an encyclopedia and books; he would give them assignments and then mentor their scholarship as often as he could, given his own busy schedule.

====

3 March 1905

I had to stop packing, if only for a moment. I am sitting in my office at the end of the hall, and every action I take is an indirect correlation, making it less and less mine, the office I was afforded finally, based on merit and my applied knowledge of electromagnetism. I have shared it with another woman, much younger, an amanuensis, though she rolled her eyes and goggled at my math's abilities.

Now, the time has come. I feel my age more each year, and this spring marks my departure from the Cavendish Laboratory.

Years spent in one place accumulate a great deal of belongings. Papers, awards, discarded bits and bobs. I have stopped to write this entry because I found "**Flyology**" tucked haphazardly between two volumes I must have hastily put away—Five years ago? Fifteen? —in the years since 1874. I traced my hand over my once beloved copy of twelve-year-old Ada Lovelace's treatise on bird wings, the book that started it all for me.

A starling squabbled just now as it passed my mullioned window. I can look down and see the melting snow in the Old Court, and I think that the English complain so easily. We had much harsher, bitterer winters in my girlhood in Scotland. We curled up fireside and told stories; we read books.

I have often wondered where I put it, "**Flyology**"—thought I misplaced it, blaming my aging memory. Today, I feel as though I am greeted by an old friend.

====

Bethan Sutherland has been my best friend since our earliest childhood. We were born and raised on adjacent dairy farms near the tiny southwest Scottish community of Corsock. A little over a two-mile hike due south of Corsock brought the hiker to Glenlair, the family estate of James Clerk Maxwell. The hiking trail connecting Corsock with Glenlair was well worn. We began visiting Mr. Maxwell's estate early in our lives.

Bethan is a little less so, but I am a curious person with a rich imagination. Each day in my wanderings with Bethan, we were almost certain to come into contact with Mr. Maxwell, who had just become Dr. Maxwell, one of Scotland's brightest young men. I was sure he would soon occupy a position at the top rank of the world's scientists. I was sure he would stand alongside Sir Isaac Newton as one of Britain's brightest minds someday.

On one particularly cold morning in my nineteenth year, I found myself in Edinburgh, shivering even while wearing my warm coat, hat, and mittens as I waited for the small coal stove in Dr. Maxwell's office to heat up. Dr. Maxwell had sent me up from his private office on his family estate at Glenlair, a day's train ride away, to do some research in the university library. I'd scrounged a copy of Edinburgh's "**Scotsman**" from a desk in the office. The tea would be hot soon.

The office had been provided to Maxwell as a courtesy by the University of Edinburgh in recognition of his graduation from the school and his likely future contributions to physics following his brilliant work as an undergraduate. He had gotten to Cambridge for his advanced degrees, as well as, more recently, King's College, London.

My parents were teachers and farmers, and our house was full of books. Our conversations at dinner were full of grand ideas, travel, and detailed mathematics. Maxwell was a personal and family friend, a teacher, and a mentor and physicist.

Remembering Maxwell's attention to me as a young girl always makes me feel happy. From my earliest memories, he always treated me like a good friend and, once I worked alongside him, very much like a professional colleague. I remember all the puzzles and games he would bring to my attention when I was younger. As I stumbled over the challenges presented by these entertainments, Maxwell would offer suggestions on how to work through the knottiest problems. For the very most difficult problems of all, we would work together. I remember these moments, especially, with great fondness.

Around the age of twelve my parents decided I had progressed much further in my learning than they could keep up with by homeschooling in Corsock. They asked for Maxwell's help in getting me enrolled in the Edinburgh Academy, the school he himself had attended.

I was not happy to remember that I experienced my first rude awakening to life's realities when Maxwell came to our home to tell us that enrollment in the Academy would not be possible. "The Academy is an all-boys school," he told us. "It is a very old tradition, and they are very protective of it."

Surely, he could see my disappointment because I could not hide it any more than I could hide my tears as he spoke the words. My parents were downcast, tears were flowing down my face, and all was quiet in the room for some moments.

Eventually, Maxwell spoke. "Let me think on this. You all, as a family, should continue to think in terms of sending Eirian to Edinburgh. I am now a senior professor at the university there, and I have complete access to all the teaching resources it has to offer. I have developed good relationships with several members of the teaching staff, and I believe I can encourage them to act as her tutors from time to time.

"For myself, I will continue to oversee Eirian's coursework, and will agree, here and now, to read and grade all of her papers. The important thing, however, is her moral upbringing. I think it is very important that you both accompany her and continue with the most important work of parenting and encouragement in her studies until she feels she has gained enough confidence in herself and awareness of the world to be on her own. You are both excellent teachers, and I can help you find appropriate employment, introductions, and housing around the university community."

He concluded by saying: "My family will take care of your properties at Corsock while you are away. You will always have your home to come back to."

***

I recall that day in Corsock as being one full of extreme emotions, even at my age of twelve years. My parents and I walked with Maxwell back to his carriage. The sun had broken through the usual clouds and drizzly rain. The rolling green hills suddenly exploded in the distance, filled with sunlight.

In Maxwell's Edinburgh office, I felt a little guilty as my mind wandered to the long walk to a stand where the horse-drawn cabs usually waited. I had not slept well in the rooming house near campus the night before because the difficulties of Maxwell's assignment were roiling my brain. I rubbed my eyes at the thought of another night with a few hours of restless sleep—if, indeed, I would enjoy any sleep at all.

I picked up the notes on the statistical problem that Dr. Maxwell asked that I bring with me so that I would be able to do the necessary research in the university library. In particular, he asked me to carefully examine his recent paper titled "**A Dynamical Theory of the Electromagnetic Field.**" I knew I would have to build my skills in partial differential equations and statistics, and I would need to gain some level of understanding of his formulations of the so-called operators "div," "grad," and "curl."

These concepts were the result of his deep thinking about the "shapes" of forces at work when concepts of electricity are joined with concepts of magnetism in ways that Michael Faraday had investigated and published in his 1859 book ***The Forces of Matter***. I knew that these concepts would be at the heart of any of Maxwell's future writings on electromagnetism.

I tried, initially without much success, to envision what these terms might mean in describing shapes occurring in a space that had no physical connection with the three-dimensional world I could see with my eyes, touch with my hands, hear with my ears, taste with my tongue and smell with my nose. Maxwell suggested that by envisioning these shapes turning freely in space I would more easily see how to apply the mathematical formulations he had developed.

He believed that Newton had developed his calculus in this way, by envisioning the shapes of the forces that connected the planets. I found this idea of experimentation occurring in the laboratory of the mind rather than in the laboratory of flasks, beakers, sinks, and gas jets profoundly difficult, but I resolved to continue until I got it right was what Maxwell wanted. I knew also that what Maxwell needed if he was to communicate his complex and not-yet-popular ideas to a less scientific public who might sit in judgement on his funding requests, or even to the more scientifically literate members among his professional colleagues who might be persuaded to support his efforts had to be a lot of care and imagination in the preparation of his supporting documentation and presentation materials; ...the story-telling in other words.

One of the many reasons Maxwell brought me into his laboratory was his appreciation for my deep curiosity, almost a match for his own, and my probing intelligence and facility with math. He also knew I was ready to get my hands dirty in the lab. He enjoyed reminding me of the time he came to visit my family when I was barely six years old. I was down in the mud trying to pull a screaming piglet out from under a pile of wood, where his tiny hoof had gotten stuck, so I could reattach the small animal to his mother's nipple.

When he told these stories in public, especially to his professional colleagues, I could not help but turn away to hide my deeply flushed face.

Maxwell's concerns about education and schooling were limitless. Whenever in Glenlair, he worked tirelessly to establish a school and church. While at Mariscal College in Aberdeen, he had established workingmen's colleges and conducted regular lectures to those who would become engineering aides, builders, mechanics, draftsmen, and administrators. In Maxwell's expansive vision these young men—and as many women as he could entice into the work despite the many obstacles to women working outside the home—would have the responsibility of assembling and maintaining the environment and structures housing the electrical and steam machines that would power the Industrial Age.

His concerns about the need for technical training across the Industrial Age were not limited to the factory floor and the machine shop. He also held regular weekly lectures and hands-on instruction for university students who were then able to build lab equipment and do the lab work and calculations in support of the work of scientists like him.

Maxwell's strength as a scientist, as yet unrecognized by many of his professional peers, lay in his refusal to fully accept ideas that could not be tested. His scientific thinking on electricity and magnetism, for example, was far ahead of most of his colleagues in that they held firm to the concept of a mechanistic, clockwork universe as shown by Sir Isaac Newton almost two hundred years before.

Maxwell's testing of the ideas of Michael Faraday, and the frequent electrical shocks he got from wires and gizmos when the testing went awry, gave him profound doubts about this dogmatically traditional view of how the universe works.

Testing Faraday's insights also meant that my mentor must devise the tests and, more often than not, design and build the test equipment in his workshop in order to do testing on new, unorthodox ideas that had never been tested before. Indeed, the equipment for testing the ideas had never been built before.

***

I looked around the little office at the pieces of testing gear that Maxwell brought with him when he left King's College, London. Of all of these pieces, I found myself most interested in his color top. This device was a step along the way to a separation and definition of the spectrum of colors contained within sunlight, a step that would lead eventually to his creation of the world's first color photograph.

This photograph would also illustrate his ideas about the electromagnetic spectrum.

Maxwell had asked me, when I had some free time, to look over the draft of his paper "**On Reciprocal Figures, Frames, and Diagrams of Forces**." I knew that this paper was a step beneath the level of his work on electricity and magnetism. Though the paper applied to his interest in architecture, and the details of building construction, I could not help but think that there was a larger message at hand. This was a method for the formulation—the "forming," if you will—of general solutions to problems involving Newton's "Action at a Distance" ideas that were at the heart of our investigations into electromagnetism.

I cannot help but think that I was very privileged to have been given this assignment. I resolved to never allow myself the luxury of a mistake when I developed my own work on this project or what came to be the many future projects I would undertake.

My working relationship with Maxwell at this time was somewhat complicated by the fact that he was living and working mostly at Glenlair, about eighty miles southwest of Edinburgh. There, he was only partly concerned with advanced mathematics. Mostly, he was engaged with his family in a plan of architectural work and development on the main buildings of the estate, which was in line with the vision of his father, John, who had died in 1856.

Occasionally, I clamber onto the train, arms full of reference materials and mathematical formulas, to bring to Glenlair because I needed Maxwell's close attention on the solution of some problem. I would also do most of my editing work there, since he was close at hand to consult on test results, wording, and presentation. He would knock on the door of my workroom at the family estate and ask, "Are you ready to go over your work with me?"

Sometimes Maxwell would return with me to Edinburgh to meet with banks, carpenters, and stonemasons involved with the expansion work at Glenlair. Sometimes he would sit through a peer review of his scientific work and discuss any changes in direction that were indicated by his most recent findings. Occasionally, I went with him to whatever social events might accompany these professional meetings.

I know that Dr. Maxwell received a lot of inquiring looks when he introduced me to his professional colleagues who might not have met his wife, Katherine. After all, there were no women enrolled in a degree program at the University of Edinburgh. His colleagues wondered why he did not choose one of the many very bright, very talented young men in physics or math from the rich academic communities in Edinburgh, or, indeed, from any of the many institutions of higher learning in Scotland.

Also, because I was made more attractive to many of those same young men because of my status with Maxwell, I saw the rather knowing looks I got from older women and men outside the scientific community. I long ago gave up being concerned by these provocative, but also, often, rude and insulting social cues.

***

Whenever in London, Maxwell took every opportunity to meet with Charles Babbage and the aging Faraday to discuss the future of science in this new industrial age—an age based on coal, the manufacture of steel, and the technology of the steam engine, and, of greatest importance to their common interests, the telegraph.

I remember a visit to the Oxford University campus in July of 1862. I and my childhood friend from Corsock, Bethan, had been invited by Maxwell to travel with him to that ancient and prestigious university to attend a summer lecture by Charles Dodgson on the mathematics of Euclid. Rumor had it, according to Maxwell, that Dodgson intended, some-day, to put the axioms of Euclid into a Shakespearean theatrical format; each axiom would start a new scene.

Maxwell thought Dodgson's idea was preposterous, but also thought the lecture would give us an interesting insight into the weird, wonderful, and magical mind of Charles Dodgson.

Dodgson was a sometime mathematician, many times a storyteller, always brilliant and totally creative in bringing far-flung ideas into his ambit as senior lecturer in mathematics at Oxford. His skills as a mathematician were not stellar, and due to his stuttering, his ability to present and lecture was erratic at best, but he fashioned wonderful and fantastic characters, figures, and situations to bring dimensions, color, smells, and life to his narratives. In the lecture we attended, he sprinkled references to hookah-smoking, talking blue caterpillars; disappearing Cheshire cats; waist coated white rabbits; and officious dodo birds with abandon.

Maxwell believed there was a connection between heaven and earth, and he had dedicated his life to finding it. Despite the

Scottish conservatism in his point of view, Maxwell sometimes let his mind wander along the pathways suggested by Dodgson's imaginative creatures and fantastic events. Maxwell had other business with Babbage and Faraday in Cam-bridge and London on this trip, so he excused himself after the Dodgson lecture to catch a train for the sixty-five-mile trek to Cambridge.

Despite his fairly clumsy presentation, Bethan and I decided to see if we could get some time face-to-face with Dodgson. We knew little about him except for the interesting stories that Maxwell had told us.

After the lecture, we approached Dodgson with a couple of carefully thought-out questions, but before we could pose them, he immediately invited us to join him and some of his friends on Oxford's rowing ponds. His friends included H.J. Liddell, dean of Christ Church, Dodgson's College; Liddell's wife, Lorina; and their daughter, Alice.

As an afterthought, Dodgson invited one of his bright young freshmen in mathematics, Sean McCabe, who had attended the lecture, to act as our guide and interpreter.

Sean was a nice young man, but he seemed very shy. When he introduced himself to Bethan and me, he stuttered as well. "H-h-hello. M-m-my name is Sean. I'm-m-m happy to meet you."

I had been working to take the hard edges off the deep Scottish brogue I inherited from my parents and our surrounding farming community, but it was still strong enough that I could not hope to fool a member of the English upper class like Sean. Because Sean stuttered, however, I hoped he would be sympathetic to my difficulties with speech and dialect on the grounds of this most traditional of English institutions.

We introduced ourselves and walked together to the boating ramps.

Later, as Sean stepped away to speak with someone, Bethan turned to me and whispered, "Sean is a very nice-looking young man. Do you think he will someday get over his stutter and his shyness?"

"Yes, I think so. Perhaps he stutters because his mentor Dodgson stutters," I replied.

As the day on the rowing pond evolved into an afternoon of tea and conversation, I took note of Dodgson's fondness for the young Alice. She was not more than ten years old, but very pretty, charming, and quite mature for her age.

"They seem quite infatuated with each other," I said to Bethan as we looked upon Dodgson and Alice. Dodgson had found a deep hole among the roots of a large tree near the stream, where rowers passed by in their punts. Alice leaned over to look into the hole, and it looked for an instant that she might fall in. Dodgson caught her by her waist and pulled her back out. She looked at him and began to giggle.

Dodgson stepped a few paces away and sat on a nearby bench. He looked toward Alice, then motioned for her to come over. He pulled a set of loose notes out of his satchel. She sat down beside him as he began to read to her the story that would become ***Alice's Adventures in Wonderland.***

"Look at the father," Bethan said. "He seems a little uncomfortable with their behavior. "

I looked in the direction of Alice's parents for a moment. "But not so much that he wants to try to break it up. I wonder what's in that manuscript."

"Maybe more strange caricatures of Mother Goose rhymes or the Brothers Grimm or Hans Christian Andersen fairy tales, like in his lectures."

I turned to my friend. "Before we decided to come down here, Dr. Maxwell told me that, in addition to the business at Cambridge, the other reason he wanted to come down to Oxford was that Dodgson had been working on a book of fantasy, and James thought I might enjoy meeting Dodgson. He told me to think of the trip as a bonus payment for the editing and testing work, I have been doing for him on the equations. I am glad you decided to come with me, Bethan."

"I'm very happy you invited me to come with you, Eirian. Thank you." We held hands for a moment, then Sean returned and suggested that they join the others for refreshments.

When I asked him about the nature of Mr. Dodgson's planned manuscript, he explained that Mr. Dodgson was writing a book of fantasy that featured the girl.

"He is concerned about her reaction to it," Sean said, "so he reads from it, to her, at every opportunity."

***

The recently published writings of Michael Faraday were also important to me at this time, though not because I was personally interested so much as because my mentor insisted that I become familiar with it. In Faraday's lab work in the 1830s, he had developed ideas and demonstrations around the concept of fields of force such as those that curve to connect the positive and negative poles of a bar magnet. These lines of force can be persuaded to show themselves by laying a piece of paper over the magnet, sprinkling some iron filings on the paper, then lifting slightly and lightly shuffling the paper while moving it around above the magnet.

During the shuffling, the filings fall into the curved lines that represent current flow—the force fields between the two poles of the bar magnet. These curved lines represent action at a distance; in other words, the action of one object transmitted to another object without an obvious connecting medium. Understanding these lines of force and the methods of transmitting force from one object to another across what appeared to be empty space were very important to all the work in electricity and magnetism being done and described by Maxwell himself.

Faraday's findings established the basis for the concept of the electromagnetic field in physics. These studies in turn had evolved into the electromagnetic spectrum that Maxwell had investigated and further developed with methods in advanced calculus while a department head at King's College.

Further development of these concepts was now Maxwell's life's work. Testing, documentation, and publication of these findings would consume much of my time over the next few years.

***

As I previously stated, Maxwell had asked me to focus on the mathematics of statistical methods. When I asked why these methods were important, he replied that he was trying to account for the actions of individual molecules in a gas.

"Obviously, such molecules, in their billions, cannot be counted or tallied like children through a turnstile," he said, "so there must be some other way that satisfies, or comes close to satisfying, the rigors of mathematical formulations without the work of some kind of tally counts that would be enormously expensive.

"I have envisioned a thought experiment where there is a chamber with a permeable membrane separating a gas that is cooler from a portion of the same gas that is warmer. Imagine there is some kind of a mechanism that acts like a gatekeeper, judging each molecule according to its temperature in comparison with the average temperature of all the cool molecules.

"If the temperature of a molecule of cool gas contrasts with the average temperature of all the molecules in a way that lets it cross the membrane, then it will have the effect of cooling the average temperature of the molecules of warm gas."

He rubbed his chin. "Of course, the gatekeeper works the same way on the warm side, except it is judging a single molecule in terms of the average temperature of all the molecules on the warm side. One of my general questions is: How would we design a statistical method to help determine within some range of error how much time must pass until the temperature of the cool side of the membrane matches the temperature of the warm side? What are the attributes of that method, of that gatekeeper?"

My mind had begun to spin. I remembered one of Dodgson's stories in the lecture hall at Oxford two years ago last July, about the blue caterpillar sitting atop a mushroom, and how he had become so pontifical. In my mind I could see the caterpillar smoking his hookah, sitting in judgment of the temperature of a particular gas molecule when compared with the average temperature of all molecules of the same gas. Somehow it seemed an appropriate image.

"It seems almost demonic," I thought aloud.

"Do you mean like a 'demon,' Eirian?" Maxwell asked, rhetorically.

I didn't respond. The answer seemed a little too close to a truth that I was not yet ready for.

Maxwell, however, moved on. "Can you craft some methods that integrate changes in temperature and pressure over time for the gases on both sides of the membrane, and then link these changes to the numbers of molecules that must pass through the membrane in order to achieve equilibrium on both sides?" he asked.

Despite my muddled head, I must have mumbled that I would do my best, as I was soon put on the case.

In later years, of course, the "gatekeeper" membrane between the warm and cold quantities of gas became known, among scientists as "Maxwell's Demon." Soon after our discussion on the movements of gas molecules, I asked Maxwell to describe his equations of electricity and magnetism and what exactly he was trying to measure.

"Of course," Maxwell replied. "I'll put them on the board for you." He hastened over to the blackboard and snatched up a bit of chalk.

"Start by thinking of things that float or move in space without any visible connection to the ground, or to any other nearby physical object," he said. "Your imaginary object could be a hot air balloon, an arrow in flight, a cloud, or a puff of smoke. The question is: what force or forces causes each of those objects to change? In the balloon you might think the gradual loss of heat would eventually cause it to collapse and fall to the ground. The arrow will eventually fall from the sky, but from what cause? The air it must push against. The speed it will lose with time as it flies.

"How about the cloud? What causes the cloud or the puff of smoke to change its shape or its color? Is it the sun? Is it changes in the cloud's moisture content? Is it the pressure of the changing wind? Actually, it is probably all of those things in some complex and unmeasurable combination."

He paused to gather his thoughts. "Think about the clouds again. This time think about the dark, windblown thunderclouds in the late spring. Of all the things that are roiling about inside the cloud, what will cause the tornado spout to drop to the ground, or the lightning to come down from the cloud to strike the ground, possibly catching something on fire, or even killing some person or animal? The power of that lightning bolt, the force of it, the heat of it, and the bright flash of light can all be measured over time using Newton's calculus, and an extension of that calculus called a 'vector,' which has measurable values for both direction and force.

"Now think of a bar magnet that you move slowly toward a piece of metal plate. Depending on the size of the magnet, you might not be able to ultimately stop it from hitting the metal and sticking to it in a way that may be stronger than your ability to pull it back.

"The power of that magnet, the force that draws it inexorably to the metal plate—like the lightning bolt, but without the flash, the noise, and the drama—can be measured over time using Newton's calculus, and the same vector extension, but with a math that works for magnets.

"To link an electric force and a magnetic force to create an electromagnetic force with a given magnitude, you must use the partial differential equations that I have developed while at King's College."

Maxwell then took me through a deep thicket of calculus. In that way he hoped to get me started on a path to a true understanding of his work. I scribbled furiously along behind him.

At the end of it he finished with: "Curl H can be read as 1 over the constant c—c, Eirian, is defined as a gearing ratio between the E and M force—times the change in E over the x, y, and z axes over time.

"I'll assume you have got all that," Maxwell said. "I'll emphasize the fundamentals once again, just in case."

He turned away from the blackboard and winked at me.

"The terms 'div,' short for divergence, and 'curl' are ways of representing how the forces E and H—the EM force I mentioned earlier—vary in the space immediately surrounding the point of our inquiry. Div is either more outwards than inwards, and is called div greater than zero, or more inwards than outwards, and is called div less than zero. Curl, on the other hand, measures the tendency of the force to curl, or loop, around the point of inquiry, and gives the direction of the axis about which it curls."

Maxwell sat down in a chair facing me. I felt more than saw this as my head was down as I quickly tried to capture all the insights into the magic of electromagnetism that Maxwell had so willingly and effectively given me over these past few minutes. When I realized he was watching me intently, I stopped my scribbling and looked at it. The page was an overwhelming maze of numbers, letters, and signs I felt I could lose myself in.

"I don't know if I will ever understand this," I said, resignedly.

"Don't feel alone in that," Maxwell said. "I don't think that either I or Faraday have a deep enough understanding of these forces to explain them well, even to those of my professional colleagues who believe, following Newton, that the universe is a finely tuned machine. Newton's comforting machine is very much like a Swiss watch—a machine with gears and winding mechanisms and hands to show the time, each connected one to the other with no empty space between.

"Faraday and I are talking about those spaces, the empty spaces where the power of these forces of electricity and magnetism reside and do their work. To proceed from a lifetime of belief in a clockwork universe to a universe where forces work at a distance with nothing connecting one object to another is hard for anyone to grasp, scientist or not.

"The introduction of these ideas of field and of actions at a distance disconnects the forces of the universe from the movements of the objects these forces are thought to influence. Our case will be a very difficult one to make in the face of such thinking, entrenched as it has become.

"Even so," he continued, "it will be your job to carefully watch and re-cord the details of my experiments with electricity and magnetism. Then, though I won't require it of you, I'd like it best of all if you would at least try to repeat them without any guidance from me, to see if you can get the same results.

"Whether you try to do your own tests or not, your final task for me on each experiment will be the preparation of text and illustrations describing each of the experiments as you saw them. Expect my very careful review because it is in these descriptions that we may all find the best way to present our findings on electromagnetism to the very large group of my skeptics in the scientific community."

And so it is that I found myself in January 1866 that I find myself working at Glenlair. Charles Ludwig Dodgson, whom Bethan and I met at Oxford in 1864, published a book titled ***Alice's Adventures***

in Wonderland. He published the book in November 1865, using the pen name Lewis Carroll. Maxwell alerted me to the publication and asked if Bethan and I would like to travel to England in March with him and his wife.

"I will be at Cambridge, but I can make travel arrangements for you both to go to the Oxford campus for a day or two," he said. "Perhaps you can meet with Dodgson there and have him autograph his book for you. Then you can come down to Cambridge to join Katherine and me for dinner with some friends of ours."

Maxwell said later that he thought he saw an actual glow in my eyes. In any case, he noted that I did seem briefly incapable of speaking.

"Oh, thank you, Dr. Maxwell," I said, once I'd recovered. "I am overwhelmed by your kind and generous offer. Are you sure?"

He smiled at me, so I did not wait for an answer.

"I must go today to discuss this with Bethan. We have so many things to prepare for before we leave!"

***

Before returning to Glenlair, Maxwell, his wife Katherine, Bethan, and I were invited to a London dinner party at the house of Maxwell's long-time mentor and friend, Michael Faraday. According to Dr. Maxwell the party would honor Benjamin Disraeli. Though not yet resolved, the collapse of the Gladstone government was virtually assured. Gladstone, a bitter enemy of Disraeli, would clearly be dispatched from office, along with his liberal party in the upcoming elections of June 1866.

Disraeli was now sixty-one years old. He was a friend of Queen Victoria, a novelist, a member of Parliament, and a continuing supporter of Tory governments and politics. He would soon be appointed chancellor of the exchequer, and Conservative leader of the House of Commons in the third minority government to be organized by Lord Derby, a Tory, as soon as he was sworn into office.

The dinner took place at Faraday's home in Hampton Court, Middlesex. Though without formal education beyond the eighth grade, the elderly Faraday had been greatly honored by his scientific peers and by Queen Victoria herself. He had been offered an opportunity to be buried at Westminster Abbey among the resting places of England's royalty, but he refused, preferring to remain with the common man even in his final resting place.

In fact, this particular get-together was Queen Victoria's idea in the hopes of exposing the conservative and sometimes prickly Disraeli to some liberal thinking in advance of the coming change of governments. The choice of other guests at the party, the queen suggested, would be up to Faraday.

Faraday had been living in Middlesex for the seven years following his retirement. When told of my visit to hear a lecture by Dodgson, Faraday insisted that I invite the man and whoever else might be good company for him. Charles Babbage, another of Faraday's guests, had recently read Dodgson's new novel. He expressed delight at the prospect of his joining us at dinner.

"Dodgson's novel is the talk of London," Babbage said. His eyes were all aglitter at the thought of Dodgson with his stumbling speech arguing with the articulate Disraeli about the parliamentary intrigues of the day. Apparently, when he thought of Disraeli parrying Dodgson's literary wit with some parliamentary and political fantasies that Disraeli would try to promote, Babbage could not restrain himself and broke into gales of laughter.

At one point, Babbage even excused himself to go to another room where he gave in to bouts of uncontrollable laughter and fits of coughing. Other members of our dinner party began to worry about his health, but he returned after a few minutes, waving away any health concerns.

"I apologize to my host for my behavior. I hope you will all forgive me for my ill-mannered behavior regarding a guest who may not be as fully articulate as others of us. Please trust me when I say that it is not the impediment that I find comical, nor the speaker, whom I admire and consider a friend and professional colleague in mathematics. The problem is that I imagined the sounds and conversational tension such a debate would carry to the rest of us. It tickled my funny bone, and my imagination got the better of me."

Though I found such a joke odious in its humor, I nodded my approval of his conciliatory remarks along with the other dinner guests. We all turned back to our conversations.

Shortly after, an aide to Faraday, acting as butler, announced the arrival of Charles Dodgson and Sean McCabe of Christ Church, Oxford. All of us rose to congratulate the new author on the publication of his very successful first novel. Though we had all been together on the Oxford campus over the past two days, Bethan and I approached Sean and asked that he join us at the table when convenient after introductions.

"Th-th-thank you," Sean said. "I will join you sh-sh-shortly."

In an aside to Bethan on our way back to the table, I noted that Babbage would not find the speech problems of the guests so entertaining if we reminded him of his many crusades against minor issues of minimal public interest.

Sean had told us of Babbage being denounced during debate in Parliament two years ago for "commencing a crusade against the popular game of tip-cat and the trundling of hoops."

The issue was a serious one having to do with interference with traffic along crowded roads, and with frightening horses pulling carriages into public rights-of-way. But the tabloid press in London has, historically, liked to bring public celebrities down to size with humiliation. They are always looking for targets. They picked up on the issue of 'trundling of hoops' from the parliamentary debate and turned it into a riotous play on language. This had caused Babbage to become a laughingstock—one that I was feeling at the time, with righteous anger on the behalf of my friend, was well deserved.

We waited patiently while Sean explained all this in his halting voice. He was very grateful to us for our kind of patience with his speaking. As a matter of fact, we were both wonderfully entertained by the way Sean manipulated his speech defect into comedic descriptions of the popular news of the day.

I privately noticed that Sean's speech improved as he spent more time with Bethan. When I asked her about it later, she gave a coy smile and said, "Why, yes. Sean's stutter seemed to go away after a while. I think the stutter is only really there as long as he is under the spell of Dodgson. He is really quiet an unusual young man."

"I am delighted to hear it," I said.

***

The next day, we Scottish travelers began our three-day journey back to Glenlair and Corsock. We all agreed that it had been a wonderful trip. The stories we could tell about all the wonderful people we met and the events we participated in would last us the rest of our lives.

I was mulling over the ideas in the book ***Alice's Adventures in Won-derland,*** though I was not sure I could see any value in them that would benefit my work for Maxwell. I did love the book, though, and all of the fantastic imagery and wordplay in it. I resolved to keep thinking about ways to link the magic of the Cheshire Cat or the Mad Hatter or the March Hare to the hard realities of Newtonian math.

Maxwell would later share with me all of the things he had been told in his business meetings at Cambridge. The university wanted him to have a leading role to play in the rapidly emerging sciences of the Industrial Age. The development of new industrial materials and sources of energy were moving fast across all of Europe, and Cambridge wanted to build a science laboratory worthy of that role.

More important to Maxwell, they recognized his lifelong dedication to all aspects of science and, in particular, his collaborative work with Faraday to perfect the scientific understanding of electromagnetism. The leaders, statesmen, and financiers putting the project together proposed to hire Maxwell to build and run it. It would be called the Cavendish Laboratory, after our nation's acclaimed chemist and physicist Henry Cavendish.

"James Clerk Maxwell will build the lab, and, once built, he will become the first Cavendish Professor of Physics," they said. "If, of course, he decides to take the job."

But first, Maxwell had an obligation to his beloved father, John, to complete the expansion of Glenlair. Other than our work together, there would not be much new science at Glenlair during the next few years.

Katherine, Maxwell's wife, an accomplished scientist in her own right and indispensable helpmate, was not at his meetings at Cambridge in 1866, but she confided in me later that she could see the faraway look in his eyes. She knew that big things were brewing, as they always were with her husband, and she would find out more about them in due course.

***

Maxwell spent the rest of his life developing the Cavendish Laboratory, particularly the pioneering physics labs that would move physics out from under the unbending intellectual influences of pure mathematicians and into the hands of experimenters like us, in materials, objects, light, and the fundamental forces of nature. I stayed in his employ the entire time, helping with the design, development, contractor management, and high-level staff recruiting through the Cavendish opening in 1874 and his appointment as first Cavendish Professor of Experimental Physics.

Maxwell died in 1879. Because of my deep association with him over so many years and my deep understanding of his work the Cavendish has kept me on until now, as a teaching and lab associate.

But I feel it in my bones. Each winter, even if it is one of these puny, balmy English ones, is harder on them. It is time to enjoy the spaces in between, while there's still time left to do so.

As I finish packing, I consider seeking out the beautiful, bluish violet stone that held a special place on Maxwell's most favored library shelf in his office, wherever that office happened to be. Perhaps the stone is just down the corridor, fallen behind a cabinet or stacked between two books, as had been my beloved "**Flyology.**"

Many visitors commented on its' rare beauty. No one had ever seen a stone like it. When asked about the origins of the stone Maxwell had to beg ignorance. It had been given to him by a neighboring family of wealthy sheep herders in Corsock in celebration of his Cavendish appointment. They had held the stone for several generations after the founder of the family's fortune had acquired it from the owner of an art studio partner of Leonardo da Vinci in Florence during the early years of the Renaissance.

That is as much as anybody knew, or knows, of the stone's provenance.

But I am sorry to say, my wish to find the stone is not to be. The stone disappeared shortly after Maxwell's death, lost, it was thought, in the general scramble to gather all of his scientific papers for safe-keeping.

END

Now I am Become Death

Fall, 1943 CE. Sarah, a bright, young physicist and mathematician, works with Richard Feynman at the Manhattan Project lab on a desert mesa called Los Alamos near Santa Fe, New Mexico. Under Feynman's loose supervision, Sarah will design the procedures, and then test the mathematical formulations needed to measure experiments with bomb architectures, container shapes, and expansion rates of bomb-blast effects.

Sarah's colleague in the Theoretical Division's Diffusion Unit is named Konstanty. He is the American-born son of a Russian émigré, who arrived with his family following the events of 1917. Konstanty's family was devoutly Russian Orthodox, and there was no place for them in the new Russia.

In order to secure his citizenship, Konstanty joined the US Army as soon as he came of age. There he learned important skills in the management and use of malleable, high-intensity explosives. He apparently had not learned enough of the most important skills because his last explosive setup blew up in the process of connecting the firing changes to the triggering device. Such accidents among those trying to rapidly learn new technical skills and scientific methods were not uncommon in the mesa's intense design, development, and testing environment.

Sarah is now faced with finding a replacement for the bedridden Konstanty. Worse, she has already been dealing with the need to fill a supervisory vacancy for the computation unit. That unit is charged with running all the explosion data through a complex series of procedures, all through the use of hand-cranked calculators, to determine whether a particular experiment has produced useful data and, if useful, what it shows about the effectiveness of a particular design of the chamber within the bomb that will trigger the explosion.

Dr. Richard Feynman, the leader of the Diffusion Unit and Sarah's immediate supervisor, has not been of much help so far. Feynman is young, and his contribution to the project is a rare combination of theoretical brilliance and practical experience with mechanical gizmos, radios, electronics, and household appliances gained in his father's Bronx repair shop. Feynman is so often called away to help untangle some critical problem in some other high-priority operating unit that he has delegated almost all of the operations of the Diffusion Unit to Sarah, including, as her first assignment, the recruitment of her two primary subordinates.

***

Several months ago, Konstanty had called out to me. "These calculations are fucked up, Sarah! Do them again."

This was shortly after I began working here on the mesa.

Shortly after his accident, I went to visit Konstanty while he was still in the coma. I remember swearing to myself, but softly. I was sorry to see him in the state he was in. Konstanty and I hold the same relative position in the Diffusion Unit, but he has always made it clear that he thinks himself smarter than me. Either that or maybe he thinks he has more pull with the project's senior managers. He claimed a friendship with Dr. Edward Teller, a hungarian physicist with bushy eyebrows and a deep accent who lost out to Bethe in the competition for the job of leader of the Diffusion Unit.

I wondered why he would brag of his Teller contact because Feynman might be a little sensitive about Teller. After a while, though, it became obvious to me that Feynman paid no attention to Teller unless, of course, Teller had a technical problem that needed his serious attention and focus.

As part of my introductions to the other science managers on the project, I met with Dr. Edward Teller in his office. Teller had also been assigned to the Theoretical Division and came to the project around the same time I did. He is a gregarious man who can be charming as long as the subject isn't about his qualifications to do higher-level work or his concerns with the capability of those who now hold the positions where that work is done. He can get a little heated on these topics, as I have learned.

I changed the subject by alluding to a small stone Teller keeps in his science library, next to his lab. The stone has a deep blue color that seems almost translucent.

"I am a secret gemologist," he says, his Hungarian accents undiminished. "I appreciate the unique qualities of the stone, but the stone is rare and there is nothing in the literature available to me that speaks of this stone or its possible origins. I have never really had time to devise and run the tests necessary to determine its chemistry, nor have I been able to determine where it came from or what its value might be."

Though he has no idea where the stone came from, Teller insists that James Clerk Maxwell once owned the stone, keeping it in the library next to his lab at the Cavendish Library, which is why Teller keeps his "pretty blue stone" on the shelf next to his lab.

I told Teller that I would love to look into the origins of the stone, but I never found the time to follow up with him. It was clear to me even then, before Konstanty's accident, that the time I would need to determine its origins and rarity was simply not going to be available to me, given how I needed to support the design and development work necessary to make the "gadget" work.

Following Konstanty's accident, almost all my time will be devoted to doing both my job and his in Field Services. I tried to read the brief scribbled notes that Feynman had given me on the math and physics problem he wants me to focus on as I started my work here. He had told me that my skills, experience, and graduate papers in math suggested to him that I "could make a major contribution to the work here."

I was very flattered and very pleased.

Feynman has never really sat down with me to discuss the overall purpose of his unit, nor his personal, particular, and specific interest in the work at hand. Feynman's skills, experience, and sheer brainpower are so beyond anyone in my experience that he is often called out to work on various technical emergencies on the Project. This is all the more impressive, as he is the leading physicist in the world for his twenty-four years of age and his very green, two-year-old PhD.

His work on the Manhattan Project, from its earliest days with Fermi in Chicago, has always been clear, and focused. He sees himself helping to lead the effort to untangle the theoretical issues and problems in the way of designing and building a deliverable atomic bomb. For years there have been rumors that German Chancellor Adolf Hitler has a bomb project underway that is similar to our Manhattan Project.

Finally, Feynman and the project's senior managers like Bethe, Oppenheimer, and Groves were admonished to build a deliverable bomb before Hitler could. They were to build a bomb powerful enough to break the back of the German military and, thus, end the war.

In one of my only significant face-to-face conversations with Feynman to date, he pointed to an organizational chart, to a box fairly high up on the chart labeled "Diffusion Unit."

"My official job is group leader for the Diffusion Unit," he explained.

"Among other things, I am responsible for the mathematics, and the de-sign, configuration, and testing of various ideas about the size and shape of explosive effects—how to manage the effects of bomb blasts, in other words. Your job is the mathematics. Let me know if you have any questions."

Feynman's chart showed branches to two smaller boxes below the Diffusion Unit box. One was marked "Computation," and the other was marked "Field Services." Computations is a small staff of mostly women—many of whom are the wives of project scientists and facilities staff—who work through the masses of data generated by a single bomb-blast test.

Field Services is a small staff of mostly men who will build the blast chambers to my own specifications. Other Field Services staffers will build and install the explosive devices, and then the firing and safety mechanisms needed to control the explosion. Finally, another group will install the sensors, telemetry, and audit procedures needed to generate and verify accurate data on the composition of blast effects.

One of my senior staff - very young, very new but very well educated, eager and ambitious - asked why there were three separate elements in the work of Field Services. "The three elements really have very little to do with project security. Each of the three elements are areas of knowledge that must be developed by separate teams because no one team can develop and test all the procedures necessary to build the gadget within the project schedule.

The key thing to understand," I continued, "is that the process of coordination, communication and data exchange between the three teams will be critical. That is my job, and your job as my subordinate. With respect, if you don't feel up to this you must tell me immediately."

The staffer nodded. He seemed chastened but resolved to do the job needed as he walked out of the room. I had to smile.

Once everything has been installed and tested for normative function, a lucky member of the Field Services staff touches off the explosives and creates a big boom, usually with lots of smoke. It is a great job, though dangerous. This real danger became the lesson we all had to learn; not just Konstanty.

I have no problem figuring out what Feynman wants to see from me. My job is the calculus, and I have to use and adapt Newton's 250 year old methods for measuring the relatively slow speeds of the planets around their orbits to now measure the speed of individual atoms moving, in their billions, in orbit, around the atomic nucleus, each near the speed of light.

From talk among some scientific people I have gotten to know around the project, I have gleaned some insight into exactly how the particles at the subatomic level would begin to merge once the fissile material reaches critical mass.

In one design, known as the implosion model, two pieces of highly radioactive plutonium-239 are jammed together in a precisely shaped chamber by a conventional explosion. The efficiency of that jamming together would also dictate the efficiency of the resulting nuclear fission and, therefore, the success of the bomb itself. In a second design, known as the gun model, a piece of highly radioactive enriched uranium-235

is literally shot at a second piece of highly radioactive enriched uranium-235.

I have been on my own in the development of the precision calculus and the physical testing of various bomb and container shapes. I learned early in the process that advanced calculus using multiple partial differential equations is fairly straightforward, but equations that must account for quantum factors and must use statistical mechanics in reporting the resulting output are almost infinitely more complicated.

The word "quantum" had not yet really entered the scientific vocabulary outside the physics community. In the seventeenth century Sir Isaac Newton had called the gravity that holds planets in orbit "action at a distance." In the twentieth century Albert Einstein did have the word quantum available, but he did not accept the implications of it. He called the quantum measurements that hold the atomic nucleus together with its family of electrons "spooky action at a distance."

When I first set up my office, I left a note for Feynman asking his advice on reference materials for the latest thinking on the mathematics of quantum physics. He replied with another scribbled note: "If you don't know Maxwell's equations, start there, and write me a summary of what you learn."

I was momentarily puzzled. Surely thinking that was eighty years old, as Maxwell's was, could not be thought of as the latest thinking in the rapidly evolving science and design of a nuclear fission process driven by wartime emergency. Nevertheless, I resolved to find the paper Maxwell had written and published on the subject.

When I challenged Feynman on how old Maxwell's work was, he insisted that I constantly read about the evolution of knowledge about the structure of the atom, its nucleus, and the dynamics within its cloud of electrons. After Faraday and Maxwell. "**Einstein's Special Theory of Rela-tivity,**" he added, "is a must.

"You could also start with my notes on Fermi's work on the pile in Chicago for an overview of how I view these interactions and how they are driving the work here. My doctoral thesis might be useful in under-standing my work on 'sum-over-paths.'" Then Feynman was gone, and I had no idea when I might see him again to ask questions about the things I would be reading and the knowledge I would be gaining.

Eventually I came to understand that detailed design on the prototype bomb itself would depend on the accuracy, perfection, and reliability in the results predicted by whatever calculus I will come up with.

For all his bullheadedness and badmouthing, Konstanty had helped me in my work. I hated to admit it, but he had been right about the calculations derived from the particular test configuration that got his hackles up. Shortly after that, Konstanty was seriously injured by a test explosion gone wrong. I had been devastated, but the pressure of the work was so intense that I had to put my personal feelings aside as I began a search for another assistant.

This search will be difficult. I have to find someone who has a lot of experience with applied math and the necessary social skills to help guide the human computers who do the computations by hand. They will need to understand the importance of these computations, which will eventually lead to something much more mysterious, and much more dangerous, than anything analysis alone can lead us to.

Out of the stack of possible candidates sent to me from the Army personnel office, Thomas' name and credential stuck out. Thomas has a recent master's degree in mathematics from Brown University. He has solid paper credentials for the work; has published a couple of articles on the application of statistical mechanics; and, though it is never certain, it looks like he can pass the intense security surrounding anything to do with the project.

Also, he appears to be willing to put up with the highly secretive, sometimes aggressive recruitment process that precedes any appointment to the Project. Sometimes that willingness on the part of the candidate, by itself, raises red flags among those involved with recruitment. When a flag is raised everything stops until the problem or question is resolved. The resolution of high-level security concerns often requires an extra push by Feynman; by his boss, Hans Bethe; and by the Project's science manager, Robert Oppenheimer.

Thomas's main problem, it seems to me, is that he is young and does not know anybody in the higher reaches of advanced math and physics in play on the project. After all, Enrico Fermi, formerly a professor of physics at the University of Florence, succeeded in conducting a sustained and controlled nuclear fission reaction in the basement under an unused squash court at the University of Chicago on December 2, 1942, only about 10 months ago. Before that, plutonium, along with uranium, the two materials to be tested for the most efficient bomb fuel, only existed in theory. The facilities for making the fuel do not yet exist in physical space.

Fermi himself, with his Jewish wife, Laura, is a living example of what is really at stake in our global war against the Axis Powers. No Italian scientist, especially not one as accomplished as Fermi, was allowed to leave Italy under the fascist regime of Mussolini, but Fermi's early work in nuclear physics was so fundamental and so important that he was award-ed the Nobel Prize for Science in 1938. The Italian government allowed him and his wife to go to Sweden to accept the prize. After the ceremony, they never again returned to Italy.

On the day Fermi's "atomic pile" in Chicago created and sustained the world's first nuclear fission reaction using uranium-235, the coded message to President Roosevelt was: "The Italian navigator has landed in the New World."

I have read through the letters of recommendation for Thomas several times. As I look at them, again, the comments of his advisors and professors still glow with enthusiasm. They have told me that Thomas has thoroughly studied James Clerk Maxwell's primary, world-changing work on electromagnetism. Thomas has written a couple of papers extending some of Maxwell's ideas in ways that could enhance our work in the Theoretical Division. Plus, in the words of one mentor, "There is no question in anybody's mind that Thomas is the smartest guy in the room."

"Even at his young age," one of his advisors said, "Thomas has written at least two very insightful papers on Maxwell's experimental work on op-tics, as well as additional monographs on electricity and magnetism, and he understands the fundamentals of math in new and creative ways."

It is these heartfelt recommendations that cause me to literally ignore all the other noise associated with his somewhat offbeat upbringing and family life.

Besides, I have such an urgent need for someone who knows enough advanced math to oversee the calculation process that I have practically begged Feynman to speak for this candidate with

Hans Bethe, and with Oppenheimer if it becomes necessary to ease Thomas past any security concerns.

I learn from his file that Thomas has one other problem, interesting and largely unprecedented since Alexander Hamilton fought alongside General George Washington in the Revolutionary War. I tell Feynman that Thomas has family in Haiti. His parents now live in New York City, and one of his parents is black. He is the great-grandson of slaves.

I am on the wooden sidewalk connecting some of the Army-style living quarters as I argue my case with Feynman. Aside from his brilliant work in theories relevant to bomb design, Feynman is a very popular guy around the Project mostly because he is always available to fix some kind of machine or pick a jammed lock that is holding up progress on another scientist's urgent project.

Feynman is initially reluctant to put a word in for Thomas. From my description, he says, it sounds like he might have to use up many, many of his precious political and organizational chits in order to bring a half-black, untested mathematician with a Caribbean accent onto the project. Worse, in Feynman's eyes, Thomas may not really know anything useful about the new mathematical minutiae we are using that is needed to keep the data that drives the project's fission design, testing, and developmental processes on track.

I intend to persist enough to keep the pressure on until Feynman quits fighting me.

Suddenly, though, Feynman makes a decision. "All right, Sarah. I'll include a meeting with Thomas during my next trip to the East Coast to consult with project scientists there."

I am delighted and somewhat flabbergasted with Feynman's initiative on my behalf. I excuse myself and go to my office to start making the arrangements by phone after working out the details with our designated security officer on the mesa. Thomas will come to New York to meet with my avatar, a "departmental superior" at the university I am "recruiting for."

When Thomas asks me what university that might be, I can only say, following instructions from Groves' security staff, that they are involved with a wartime project that is top secret. Thomas accepts this idea and agrees to wait for the departmental superior to show up.

I tell him that that person will simply sit down at his table at a well-known restaurant in midtown Manhattan and introduce himself as Dr. George Brown.

The "well-known restaurant" and the details of when and which table will be made known to Thomas within two days of the meeting. His travel and hotel arrangements will be taken care of. He will only need to come to the designated ticket window at the railroad station in downtown Providence, Rhode Island, and introduce himself as Thomas Brown from New Orleans. The ticket agent will then hand him his already paid-for train and hotel ticket package.

I found out later that, in the back of Thomas' mind, he could not entirely suppress the idea that this was all a huge joke. If he let his imagination flow freely, he could see someone in a plain, dark brown overcoat jumping out of the bushes at him in New York, threatening him with a handgun that popped a flag out of the barrel marked BANG! He chuckled at the notion.

***

Sometimes, I get bogged down in a tangled pursuit of solutions to one of the problems typical of Feynman's requests. During my advanced courses in calculus, I had learned to recognize the point in the pursuit of a solution when I needed to stop trying to force a path to an answer and take a mental break. Everybody doing this kind of work has to know how to do this. I would get up from my worktable and walk around, or I would lie down on the couch and take a nap, or I would lean back in my chair and begin a process of woolgathering over things that I knew to be more pleasant.

I have often recalled the day—Friday, April 30, 1943—that I arrived at the train station in Lamy, New Mexico, after a three-day trip from Chicago. Lamy, a tiny town with a big railway station and a big hotel, the Ortiz, was as close as the Atchison, Topeka, and Santa Fe railroad ever got to New Mexico's state capital. The engineers who laid out the route realized that the hills around Santa Fe made it impossible to run the railway directly into the city.

A small spur line had been extended in 1896, and many new arrivals like me took this rail spur the fourteen miles from Lamy into the small station in Santa Fe. Feynman wanted me to get the best possible treatment, such as it was, on my way to the mesa with its hectic, gritty, muddy world and its barracks-style buildings known to the civilian and military inhabitants as Los Alamos.

I was approached by a civilian driver on the station platform, who greeted me by name. I did not know the driver, but I had been thoroughly briefed on the security protocols and procedures on the Project. I knew a strange man in civilian clothes would approach me, speak my first name, and take my bag to a car parked nearby. Once in the car, I was then driven to a pleasant residential compound at 109 East Palace Drive near downtown Santa Fe.

The entry to the compound was marked by a small, blue sign with red lettering that read, enigmatically, U.S. ENG on the top line, and then, on the bottom: RS. This was the only point of contact in the city for almost all civilian and military personnel, and for much of the material passing through Santa Fe on the way to Los Alamos.

Once inside the compound, I was taken to a small office in the back. There, Dorothy McKibbin greeted me warmly.

"I think you must be Sarah," Dorothy said with a smile. "Welcome to Santa Fe."

“Thank you,” I responded. “Here is a copy of the orders that I think I am supposed to give you.”

“Yes, thank you. You will have to fill out this form so I can give you a letter of authority to go up to Los Alamos.”

I had just come out of the warm and rather carefree embrace of academia. I was not used to the possible demand that I “fill out” an official-looking—not to mention officious and scary-looking—military form.

Jokingly, I asked, “What happens if I don’t fill out the form?”

Then I can’t issue your pass,” Dorothy responded with a smile that suddenly looked a little less warm. "I’ll let you think about what might happen on a top secret military base if you are caught without document-ed authorization for your right to be here.”

I felt admonished. I took the form, looked Dorothy in the eye, and apologized for my devil-may-care attitude. I took my time so that I could carefully fill it out.

As I turned to go back outside to find my driver and my ride up to the mesa, Dorothy left me with this parting comment: “I think you will do fine. Be glad that you had a more or less friendly introduction to project security. Up on the mesa there are some people in uniform with guns, and they can sometimes get very excited when they think there is a security problem standing in front of them. Be careful and good luck.”

I waved and thanked Dorothy. Obviously, she carried a lot of weight on the project. I expected that I would see her again. When that happens, I will know better than to ask a dumb question.

***

By late summer, processing the huge quantities of data emerging from testing designs and configurations has become the single biggest bottle-neck in the work of the Theoretical Division. Each configuration test run against protocol produces hundreds of thousands of data points; all have to be processed by hundreds of people called computers or calculators using mechanical calculating machines sometimes called "comptometers." These are made by Marchant and are powered by manual hand cranks, which gets tiresome after a while. Machine operators had to take frequent breaks to rest their tired arms, shoulders and hands.

Processing the data on the manual Marchant machines requires a series of steps in complicated but rigorous sequences. A single dataset often has to be entered into one of the machines, processed to a result, and then tested to make sure the data is "normative." If it is, then the data in the result can be entered into a different machine with a different set of steps and a different set of normative parameters.

These complicated sequences are prone to a lot of errors, which is a source of tooth-grinding frustration for me. Results have to be constantly checked and rechecked. The constant pressure to do careful, detailed work and to produce accurate, verifiable results is intense. We limit our computers' work shifts to four hours. Otherwise, they will wear out too fast, making the error counts escalate and the process downtime accumulate to unacceptable levels. Some of the staff have already started asking for transfers to any other unit of the project.

Feynman's talent for fixing broken machinery is constantly on call by the computers. He responds quickly because the results they produce are critical to the overall performance of the Diffusion Unit. He knows he can't allow this backlog situation to continue, if for no other reason than that Bethe feels the machine maintenance is taking away too much of Feynman's time, which is needed for work on the theory, design, and development of the bomb itself.

Eventually, I was told that Bethe, for that very reason, has told Feynman to stop doing machine maintenance. His valuable time and talent are needed on theory, not mechanism. In a way Bethe's concern mirrors the general concern across physics since Newton,

Faraday, Maxwell, and Ein-stein: "Is there such a thing as a 'real' object, with mass, dimensions, and boundaries that can be measured by numbers that can be added, subtracted, and in other ways 'calculated,' or not?"

Making progress in finding an answer to that profound and persistent post-Einsteinian question is very much within the mandate of the Theoretical Division and, therefore, very much within the direct and explicit responsibilities of Bethe and Feynman. And they look at me with the expectation that I know how to set up the math. I hope every day that I will not prove them wrong.

As the supervisor of the computation unit I have to think about all these things and try to follow Feynman's casual and seemingly off-the-wall comments on what I should be looking for in a subordinate. One of his comments, for example, had to do with George Babbage and Ada Lovelace. I had read about Babbage in school and some poetry by Lord Byron. I had not learned, however, that Byron's daughter Ada was some-thing of a math whiz.

Feynman had said: "When you are thinking up questions for your prospective computation supervisor, why don't you ask them what they know about Babbage and Lovelace? If they say, 'Who?' My advice to you would be to thank them for their time and send them home."

In order to avoid embarrassment and a possible firing of my own, I quietly looked up Babbage and Lovelace in the compound's library. Early in the nineteenth century, George Babbage and his eventual protégé, Ada Lovelace, had set out to build a machine that could do arithmetic functions with speed and accuracy. With extensive financial support by the British government under Queen Victoria and her husband, science enthusiast Prince Albert, Babbage and Lovelace made great strides in developing the actual machinery and programming techniques for scientific calculations.

These were to be working machines driven by steam. Unfortunately, Babbage was only able to build a small working prototype before the escalating costs of development approached that of a major warship for the British Navy in that day. The main problem seemed to be not with Babbage, but with the lack of well-developed standards of manufacturing for metal. In particular the irregular quality of brass used by Babbage caused errors that accumulated as the mechanism in the Babbage machine progressed toward a solution.

The accumulated errors too often stopped a complicated calculation procedure before it could produce a meaningful result. Babbage's inability to move his machine past the prototype stage caused the British government to terminate the development funding. Nobody stepped forward to take over the ideas in the prototype established by Babbage. Lovelace, though she was willing to continue the work toward an analytical engine, began to sicken with a gastrointestinal disease. She died in 1852 at the age of thirty-seven.

In reading up on them I discovered that Ada Lovelace was much more than a simple math whiz. Working with Babbage on what Babbage called his "difference engine," Lovelace had begun designing what was eventually called an "analytical engine" to be built following their expected success with the difference engine.

In those days weavers could weave their spun yarns into customized designs by use of a Jacquard loom. Lovelace developed a system of programming, based on the weaving methods used by the Jacquard loom, which would make it possible to analyze any data on any subject using the computation capability of Babbage's difference engine.

With the story of Babbage and Lovelace in mind, our project managers reason that, as the largest and most well-known provider of commercial data processing services, International Business Machines should be able to help us with our growing calculation problem.

Unfortunately, the problem in the way of finding answers to the technical and mathematical questions arising out of studies of nuclear fission is not only about the money. It is also about the application of mathematics in ways that go far beyond anything that has ever been tried before. The calculus developed by Newton -with important contributions by Leibnitz in Germany - in the seventeenth century can provide rigorous approaches to many of the known nuclear forces that need measurement, but there are so many unknown aspects that can only be understood by brute calculations of billions of nuclear particles moving among each other, with all their accumulated and explosively expanding masses, at speeds approaching the speed of light.

In the explosion contemplated for the bomb, the time between achieving critical mass and the maximum explosive yield is a matter of a few billionths of a second. A few billionths too early and the fission will stop without exploding; a few billions too late and the effects will be unpredictable, possibly catastrophic to an unknowable extent.

The factors affecting the length of the time between a failed reaction and a possibly catastrophic one include the mass and shape of each of the two chunks of plutonium to be merged; the shape of the merged chunks at critical mass; the shape of the container containing the merged chunks; and the purity of each of the materials in the explosion chamber.

This thinking leaves aside the question of bomb delivery. Can the device be shrunk enough, and can an airplane be built big enough to carry the bomb to its target?

As acting supervisor of the computation unit I felt it was my duty to see for myself how this process of taking raw explosion data and running it through a series of independent data processing and validation procedures worked to produce information that we could use in bomb configuration design.

***

Unfortunately for the managers of the Manhattan Project, the technology needed to extend the Babbage-Lovelace prototype into a production-capable calculation system had never been built. Whatever happened when Babbage and Lovelace demonstrated their prototype to Queen Victoria, no further development money would ever come to Babbage-Lovelace from the British government.

The Project's military and scientific managers had no real choice but to create a formal Diffusion Unit with Feynman in charge to try and figure out a solution. Among the Diffusion Unit staff we feel there is no one more qualified to build a diffusion process than Feynman.

Having seen his skills up close it has become clear to me that Feynman is the guy who will try anything, including the dumbest and nuttiest sounding ideas, even when the opposition is hostile, in his attempt to try to make something work. The result of his wildly creative but often scattershot approach to problem solving often succeeds when others have given up long before.

***

At a party on the mesa after Feynman returned from New York, he gave us a summary of his first meeting with Thomas.

Feynman met with Thomas at a small restaurant called Julio's Bistro on the corner of Fiftieth and Seventh Avenue. He recognized Thomas from the pictures I had shown him. He told me that he walked up to the man, sat down, and said, "Hello Thomas. My name is Fred Brown."

Thomas later noted that "Fred Brown" wasn't much older than he was. In fact it looked like he might be younger. "Should I shake hands with you?" Thomas asked.

Feynman's finely tuned ear could not help but notice Thomas' careful but not quite successful attempts to control his Caribbean accent. "Of course." Feynman stuck out his hand. "We're pretty secure here, but I can't tell you exactly how secure. If you come to work for us you will begin to understand the full extent of it. I understand you are from the Caribbean, and you know something about math. Is that right?"

"My father was born and raised on Haiti. His parents, my grandparents, were both black at least for a few generations back. Few people in the Caribbean will hazard a guess about bloodlines if their lineage has origins among slaves working the plantations. Whatever the bloodlines, my father became an accountant, and he was successful enough to be recruited by a major US firm, then relocated to New York, where he met and married my mom.

"My mom was a math teacher, and she and my father both spent a lot of time with me on my homework. Now, I like to think that I know quite a lot about mathematics, especially the application of some math in the area of statistics that has never been attempted before. I don't know what you can tell me, but whatever you can tell me about your current problems in computation, I can certainly tell you what I know and have experience with."

Feynman paused. "I've read your papers on Maxwell's work in optics and electromagnetism. We are finding that some of the mysteries he and Faraday exposed at the molecular level are still mysteries today, but they are mysteries now more at the level of the atomic nucleus and surrounding cloud of atoms."

"I agree with what you say about Maxwell and Faraday," Thomas said, "but what can you tell me about the math and computation problems you are having at the subatomic level?"

Feynman stared at Thomas for several moments. Thomas was calm under this close scrutiny. "Do you know any magic tricks?" Feynman asked.

Now it was Thomas' turn to stare. "Is this a question about whether or not I believe in black magic; whether or not I believe in the scientific method; or whether or not I might prefer to believe in voodoo over mathematics?"

He frowned as though he had been profoundly offended. Then he looked up at Feynman to see if he could read his intent.

"It's a simple question," Feynman said. "Give me your honest answer so we can move on."

Thomas thought for a moment, then raised both his hands in front of his face and waggled his fingers at Feynman. "Booga-booga."

Feynman's bursts of raucous laughter could be heard across the room.

At Julio's Feynman learned that Thomas also played music, the trumpet. He suggested an uptown jaunt to the Braddock Hotel, where they could listen to the house band play its signature Caribbean music.

They never made it to the Braddock, Feynman told us, and a shadow came over his face. The day was Monday, August 2. On Sunday, the day before, a black GI had seen a white police officer trying to arrest a black woman for disorderly conduct. The GI sought to intervene. A scuffle ensued, and the policeman apparently shot the GI. A riot began that lasted for two days.

***

"Focus on the question at hand," Feynman would say. "Ignore all of the issues and history that brought us to this particular problem in the first place."

This was Feynman's entire life and his entire attitude toward the problem's others had long since given up on. His doctoral thesis merely reflected his most fundamental attitudes. Paraphrasing again: "When confronting a problem, ignore everything that has nothing to do with it, and be alert to all the possibilities for fixing and repurposing. Don't hesitate to try off-the-wall fixes that might sound stupid at the time.

"Most important of all," Feynman instructed us, "try not to make a mess."

***

A month after the incident at the Braddock Hotel in Harlem, Feynman calls me, Thomas, and Konstanty to a short meeting in one of the labs used by the Theoretical Division.

Though still somewhat shaky on his feet, Konstanty is back at work and trying his best to reconnect with the other members of his field team and his other collaborators on the Diffusion Unit team.

“I assume you’ve met Thomas, Sarah?” Feynman asks.

“Yes. Of course.” I turn to Thomas. “I have been showing him around and introducing him to the human computers while we wait for our new computing machines to show up. Maybe he has some thoughts he would like to share with us about what he has seen so far?”

Thomas clears his throat. “Yeah, thanks to you both, very much, for bringing me on to this project. I have been getting the whole story from the women who are working the Marchant machines about the kind of data you are processing, and the steps you have set up to handle the calculations. My compliments on your process arrangements for handling the data. It should all go quickly from paper onto the punch cards. If we can get those machines set up quickly, we should be able to start processing real cycles of digital data within a couple of months.

“I think I do have a few mathematical tricks I can contribute to the solution of the main problem you are dealing with now, the shape of the blast effects. You will have to merge the partial differential equations of Maxwell with the statistical mechanics that can deal with the fuzzy positions of nuclear particles. Have I got that right, more or less?”

“My compliments to you, Thomas,” I say. “You have picked up a lot in a few days.”

Feynman agrees “I’m looking forward to seeing your work once you get your teeth into some of the real problems we’re dealing with here. One more thing I want you to take a look at before you start locking yourself into approaches and methods that you think will work: read my doctoral thesis, and work through enough of the math to get an understanding of how I think we need to work out the identification and analysis of forces when particles collide in a condition of critical mass."

"I'll get right on it," Thomas says.

Konstanty chooses that moment to jump into the discussion, glaring at Thomas. "Wait a minute. With all due respect, who the fuck are you? I don't remember seeing your name on any paperwork." He looks toward Feynman and me. "Did you guys forget that I work here? What the fuck is going on? Are you guys just hiring off the streets now? What is his security clearance?"

I do my best to hide my frustration. "I had to make a lot of decisions in your absence, Konstanty, including the recruitment and hiring of Thomas. There was no way to know your real condition, nor could we know if you would be back on the job in the foreseeable future. Dr. Feynman and I consider Thomas to be a capable addition to our group. We both expect him to do very important work in the integration of our manual data processing with the IBM systems that are being installed and tested as we speak."

I cross my arms. "Frankly, Konstanty, I am a little concerned about your attitude toward Thomas. I assume you have never met him, so I am wondering if you have ever had to work with a black man before?"

"No. And I don't particularly want to start now."

"If that is what you want, we can arrange that," I say. "You need to understand that I am not going to let Thomas go without a fight, let alone on your say so. I need his skills on my team. Besides, after the fights I have had with some of my sisters and faculty members in college, I don't think you are going to be much of a problem."

There is now a pause in the discussion. Konstanty is looking down at the floor. Thomas looks like he is about to say something. I scowl at Konstanty. At this moment, in my mind, he has no right to free speech.

Feynman speaks up. "You need to say something, Konstanty. Right now, I am wondering if we have a team that is capable of the kind of collaboration I need to help me get through the work in front of me. So, answer the question: do I have a team or not?"

"You have a team, Dr. Feynman," Konstanty says, "but I am not sure I should continue to be a member of it. My parents are typical Russian peasants; they suspect everybody who does not swear to the Russian Orthodox faith, and their obeisance to the practice of that faith has not mellowed during their time in America."

Konstanty walks over to Thomas with his right hand extended. "I'm sorry for my bad behavior," he says. "I am sure you are a good man, and I'm sure you are well qualified to do the work Dr. Feynman and Sarah have in mind for you. My parents have given me a load of bad impressions to carry with me in life about the evil done by people who have not lived by The Book.

"I have only recently begun to realize that almost every one of those evil people my parents talk about is of a different race than are the white European 'true believers' in the Russian Orthodox Church. I am sorry."

Thomas clasps Konstanty's hand in his. "You don't need to apologize to me, but I have to admit that I have had warmer receptions on meeting someone new. On the other hand, I have had much worse receptions, but none of those perpetrators has ever stepped forward to shake my hand afterwards. I hope you will stay on with us because I look forward to working with you."

"Sarah, I think this is a good time for you to take Thomas and Konstanty down to the explosives shack and introduce everybody around to the sapper crew down there. Once you're done, I'd like to have Thomas give me his thoughts on what we need to do to speed up the turnaround on explosion data once we get our computers set up," Feynman says.

I nod. "I'll call up a jeep. Konstanty, are you up to reintroducing your-self to your crew?"

"Yes, I am. But I want you and Dr. Feynman, and Thomas, to know that I still need to assess what I am now seeing as my new life since the accident and the coma. I still need to decide if I am in the place I need to be and want to be."

Feynman spoke up. "While you consider your options, Konstanty, I have some news you might all find interesting. Some of part of the scientific staff here have been working to resolve the technical problems of manufacturing Plutonium in the quantities we need to even run an initial test of our designs for the implosion model of the bomb chamber. We were told that the process designers for the making of plutonium might be able to produce a few grains as samples last spring. However, that proved to be way too optimistic."

Feynman continued. "In fact, they have only this past week managed to produce a couple of tiny specs of U-239. As part of my job to keep track of the manufacturing process I was able to get one of the techs involved with the process prototype at Hanford to send me these.

Feynman pulled a small box out of his pocket. It looked like a box for an engagement ring and perhaps it had been that exact thing once perhaps in making a proposal of marriage to his wife, Arline, now bedridden in Albuquerque with tuberculosis. By one barely legal subterfuge or another Feynman has managed to escape the mesa almost every weekend in order to visit his dying wife.

He flipped the lid up and with a pair of tweezers he lifted aside one corner of a piece of black cloth.

Inside, laying on the black backing lay two tiny, barely perceptible dots of something that looked like they might have been carefully scraped off a piece of light grey metal. Perhaps they had.

"Behold. Plutonium," Feynman said. "Don't touch."

I got closer. I had to take my glasses off and move them toward and away from the two particles in order to bring them into some kind of minimal focus. Suddenly my flesh grew cold and I could not stop shaking. My mind became flooded with the horrible images my family pass along to me almost every week from what they hear of persecutions of our close relatives in Europe; about Kristallnacht, and about neighbors and friends being stripped of their possessions then taken away to work camps. In 1937 my parents arranged a trip to Paris with some school friends to celebrate my good grades. My mind now flies to Pablo Picasso's Guernica hanging in the pavilion at the Paris International Exhibition shortly before Picasso had the painting taken away to the United States to keep it away from Spain's Fascist dictator Francisco Franco.

Even so my vision of the painting and of the people and animals being torn apart and the awful screaming of the children as Hitler's bombs exploded gripped me in the terrible embrace that Picasso's work had created and held inviolate in my now disordered mind. At Franco's invitation German Chancellor Hitler had done his bombing of Guernica, a small Basque farming community in northwest Spain, just to give his pilots and bombardiers and new bomb-carrying aircraft some practice dropping high explosives on helpless civilians. No warning had been given.

I backed away. I was overwhelmed with emotion. Tears came to my eyes. It was our job to take these tiny specs, these two tiny particles, and apply sciences to them that we barely understand in order to create a destructive device that we can barely imagine to burn and destroy an enemy and the men and women in their armies and navies that we have never met.

"And what of the children?" I thought to myself. "What of all of the little children?" My tears now flowed more freely. I excused myself and stepped from the room.

The last thing in my mind before I ran into the bright sunshine outside the building were the words of Alexander Graham Bell on the first successful test of a device that would become a telephone in every American home: "What hath God wrought."

It has become so easy to ignore what we're doing, holed up in our offices, distracted by the day-to-day minutiae of finding computers and hiring staff. I am ashamed that I have lost track of the gravity of our situation. We are doing something that will change the twentieth century beyond recognition. The dark spaces between atoms will strike like flint, become weaponized. They will mushroom, and light the air afire. They will rain down on children. They will enforce pain and suffering. They will support the war machine. They will enforce peace.

I look around from face to face. There is grim hope, there is fearful expectation. There is a fevered sense of discovery.

END

CHAOS

Spring, 1967 CE. Rafaela Barardi, trained in medicine to become an ER doctor, instead goes to Vietnam as an army nurse to find her cousin, an Army Green Beret. He is on a confidential mission and out of contact with his headquarters. Rafaela fears he may be missing in action somewhere in the highlands near the DMZ. Meanwhile, Aleck Morris, a reporter for a Midwestern magazine, gets approval to cover the conflict from within "the shit." When he winds up in a military hospital, he soon becomes involved in his nurse's mission to find her cousin, no matter what.

====

RAFAELA

I was kind of a tough kid growing up on my family's farm in northern Ohio. For me, according to my parents, the terrible twos were more like the terrible ones, twos, and threes. As an infant I didn't mind screaming at the top of my lungs especially if I thought my parents were watching. As I grew older, particularly when I tried to speak in sentences, my frustration seemed to grow exponentially with each failed attempt at making my wishes known by simply speaking them. On more than one occasion my folks thought something was wrong with me. They planned to take me to a doctor, but that would have required an expedition to get to a city large enough to support a doctor who might know something about my problem ...if there was a problem.

The work on the farm, particularly milking cows twice each day every day year-round, usually trumped the kind of planning and dreaming necessary to take trips of any kind, never mind trips for important business like children's health.

My parents, Marie and Angelo Barardi, are descendants of a family that emigrated from Italy in the late 1800s following the consolidation of Italy's northern and southern provinces. Unlike many of the Italian immigrants of that time the Barardis chose to settle inland on small farms miles away from the ports in the bigger cities.

I grew up the eldest of three children. Though noisy as an infant and toddler I mostly kept to myself as I grew older. I had learned how to curb my expressiveness, not so much because of the irritation it caused others, but because I was growing more self-confident about my understanding of the world and the people in it. I began to realize that few of the people around me had answers to the questions I wanted to ask.

In school some of my classmates made the mistake of thinking I was a little slow. I worked hard, and I studied hard. I was a voracious reader. At home I was a full participant in the work of our three-hundred-acre farm, where the primary cash crops were corn, wheat, soybeans, and clover. Growing required a lot of shoveling and a lot of farm equipment that could not break down when it was needed most. From an early age I took pride in my ability and knowledge of tools to help keep the equipment in good working order.

There were cows and goats for milk, a couple of pigs for meat, and chickens for eggs on the farm. There was the usual complement of dogs and cats running around; there was a single swaybacked horse for us to ride and learn to take care of. I took a particular interest in the good health, good feeding, and good care of all of the animals.

But I had resolved early in high school that I would find a way to escape the small farm town life of northern Ohio. I knew my problem: I had read too many books of too many different kinds on too many different subjects.

When I think of this time in my life, I am reminded of Shakespeare's Julius Caesar. In an aside to Marc Antony, Julius Caesar speaks of one of the conspirators: "Yon Cassius has a lean and hungry look. He thinks too much! Such men are dangerous." The line resonates constantly in the back of my mind.

My cousin, another Barardi family offspring named Rafael, also felt the early need for escape and adventure. We are nearly the same age, and we were very close growing up, more like brother and sister than like cousins. During family gatherings we often left the adults and younger children to find more interesting adventures. Eventually some of the adults would notice our disappearance. There would be momentary panic as they went looking for us, and great relief once we were found. Once found there were always hugs and kisses for the two of us errant children. Occasionally, punishment would come later for once again causing so much fear and consternation among the gathered friends and family.

One early incident in particular raised concerns about us to a very high level. It was Easter Sunday. After attending church our families had gathered to eat and reconnect with family and friends who had traveled from New York City to celebrate with their rural Ohio relatives.

The house was surrounded by woods. We had not yet sat down for dinner when the cry rang out: “Where is Rafaela?”

My father, Angelo, looked toward his brother, Rafael’s father, Fabio, his concern barely hidden.

It was Rafael who later found me, hiding in the hayloft in the barn.

Rafael resolved to join the army as soon as he turned seventeen. Our fathers were both disappointed that Rafael wanted to leave the farm. They felt he was a strong, intelligent, and practical boy who would be a great asset to the whole community if only he could be persuaded to stay. Even worse in the mind of my father, Rafael kept badgering me to start applying to medical school instead of talking about it all the time.

In spring 1960 Rafael turned seventeen. He had already advised his draft board that he would sign up as soon as he came of age so long as he could pick his service and military occupational specialty (MOS). He wanted to be an airborne Ranger, and he wanted to do basic training at Fort Ord, California. His draft board approved the request.

My high school class of twelve students had graduated in the spring of 1959. I had earned straight As all through high school. I was class valedictorian, though I declined the opportunity to make a graduation speech. Rafael and I had been discussing plans to travel together to California. There I would take some summer classes prior to entering the University of California at Berkeley for premed studies in the fall. We were eager to get on the road as soon as possible.

***

ALECK

My magazine, The Prairie Observer, sent me into Vietnam a few months ago to get what you might call “local color” on the war. The magazine and staff like me do not see ourselves as “muckrakers,” nor are we particularly biased politically, though we do try to see that deserving workers get a fair deal from the businesses that earn a profit off of their work. The thing is that we love writing. We love the process of journalism when it comes to finding people who know something about some difficult aspect of our assigned topic. We know how to arrange interviews and ask the questions that might uncover a story of interest to a wider audience.

The pay is crap and the work intense and never-ending, but we can’t think of anything better, or more fun, to do with our time in these days. Besides, the managing editor usually lets us follow our noses in finding new stories and in working those stories once assigned.

Best of all, the prettiest of the editorial assistants, Penny Blue, is, like me, a graduate in English from the University of Minnesota. She seems to enjoy my take on my reporting assignments. Penny’s editorial re-writes of my work also help keep me clear, mostly, of problems with the editorial policies of the magazine's financiers and senior managers.

So far as we know from the national news, government spokesmen, and wire-service chatter, the communists based in the north Vietnamese city of Hanoi are trying to take over the legitimate government of the south based in Saigon. The most consistent policy line out of the White House of Lyndon B. Johnson has always been, up till now, that the North Vietnamese Armies (the NVA) are mostly rice farmers who would not be able to sustain their war missions in the face of a response by the well-equipped and well-trained US military.

In late October 1965, the battle for the Ia Drang Valley, some two hundred miles out in the boonies north of Saigon, changed all that. The policy line has not changed all that much, but the four-day battle caused some high-level people stateside to wake up to the difficulty of winning this war. My magazine decided to try to get ahead of it so they could cash in on contacts and local knowledge as the war expanded.

We had been doing wire service "rip-and-read" news with press releases from Washington's official spokesmen on war matters. We figured it was time to become more relevant, and we wanted to know more. We realized that we were missing the real news of the conflict.

So, they sent me over to check it out. They did not send me so much be-cause of my extraordinary talent. I am still pretty green on the job. I think I do have enough talent to keep my facts straight, and I have the skills and experience necessary to write a simple sentence. I also have an interest in the history being built in this remote country. I even wrote a paper on it for my senior thesis that the magazine's editors considered "insightful."

Realistically, though, my assignment probably happened because I am single and free of possible complications for the magazine if something happens to me down there in the deep, deep jungle boondocks.

The conflict in the Ia Drang Valley brought four days of intense battle by one thousand US Army regulars, draftees, and specialized combat units using new, helicopter-based airmobile tactics against about twenty-five hundred ground-bound NVA troops. Even without air-mobile assets, the NVA proved themselves to be highly skilled, dedicated, well organized, and well equipped.

After Ia Drang all the earlier happy talk in the US media of an army of disorganized peasants and rice farmers began to shift away from this blind naivete. The NVA—though "defeated" in the Ia Drang, according to the news—has shown itself to be a more capable and dedicated military force than the North Koreans had been 15 years before.

There is some irony here. Like the north Vietnamese the North Koreans were a peasant army, but they had the full support of regular Chinese army troops and equipment with them on the field of battle. By contrast the Vietnamese have been in a battle of independence from the Chinese for more than a thousand years. The south Vietnamese are siding with the US military and fought alongside us in the Ia Drang. The NVA, on the other hand, is relying entirely on their own manpower and equipment.

Speaking now from the safety of hindsight I believe some of the US military planners who planned the confrontation in the Ia Drang probably now wish they had paid more attention to the thousand-year history of Vietnam's war of independence from China.

There will be a lot of American casualties, a lot of politically unpleasant consequences, and a lot of bad press coverage. Whether found to be good or bad, I had resolved to make my mark on history by contributing my full share to the news coverage.

Because of the Ia Drang, the sudden calls for an expanded draft grew louder by the early spring of 1966. My draft number was high enough that I stood very little chance of getting called. By the time I graduated from college in May, I had decided that I was not going to volunteer for military service. I was not much interested in spending a couple of years being ordered to shoot people I did not know, or being ordered to get shot at by them.

On the other hand, I am not a pacifist and had no intention of going to Canada.

The worst thing about military service, for me, was the idea that I might get shot at, hit, and crippled for life. If it happens while I'm on assignment over here, I know I will feel even worse—very angry, actually—if my crip-pling is the result of decisions by people who start sending all our bombs, bullets, napalm, and armor plate to make war on people who live in mud huts and tin shacks, who make their roofs out of tree branches and palm leaves, and who catch their food with nets they make by hand out of twigs and vines.

Even with these troubling thoughts I still felt pulled by the romance and adventure of war. As terrible as we can imagine war to be, it still appeals to the imagination of younger guys like me who are filled more with the raging testosterone of youth than with the reluctance and caution of maturity and responsibility.

Following my editor's instructions on what I should do when I arrived in Saigon, I spent the early weeks of my tour sitting at the rooftop bar of the Continental Hotel getting "acclimated." That meant sitting around listening to the same correspondents get drunk every night and talk about their time "in the shit." This got pretty tiresome. The more I listened to them, the more I became convinced that the loudest talkers had never been anywhere near the shit.

In early '67, Saigon was a long way from any combat "shit," though there were plenty of kids on scooters selling, or stealing, everything that a drunken, stupid, or horny GI might want to buy, carry, or sleep with. The shit was in the streets of Saigon as far as I was concerned. Eventually, I decided to take my correspondent's credentials and find a way out into the boonies to check out some of the real action.

I did have one interesting encounter with an Inuit GI from, I think he said, Point Hope, Alaska. He said he was a swift boat driver down in the Mekong Delta. He was up here in the big city for some R&R. He had a short, stocky body and almond-shaped eyes in his round face.

"My people eat a lot of muktuk," he told me. "Muktuk is a chunk of the skin of a humpback whale with a bunch of fat attached. The muktuk plus a lot of seal oil and caribou meat gives us Eskimos a lot of body fat so we can survive the winter." He fanned himself then. "So it's goddamn hot for me here in the jungle. A couple of my buds grew up on the North Slope of Alaska. We were all in basic at Ord together. They couldn't take the heat and the army sent 'em home. I'm still here, but it is fuckin' hot, man. Maybe I'll lucky and get sent home. That would be great because I miss my girl Mary. I need to see her soon."

We got to know each other a bit. He was quick to laugh at my jokes and I at his. At one point he took out a small, beaded pouch of reindeer hide. There was a thumb-sized blue/violet stone inside. The stone was beautiful. When I asked what it was, he said he got it a few years ago before he got drafted from some scientist working at the mouth of a creek south of Point Hope. They were preparing for some kind of atomic explosion.

"He said that the explosion was going to make it possible for us to someday have green grass around our houses for our pigs and chickens." He broke into a big belly laugh at this idea. "Pigs and chickens, he said. I laughed my ass off. I tried to tell him we didn't have any pigs or chickens; we have muktuk, seal meat, and blueberries and quail eggs in the summertime. He didn't know what I was talking about, so he gave me this stone as some kind of booby prize. I am keeping it for my girlfriend, Mary, while she is at school in Fairbanks."

Then I remembered that I did know something about the stone. The Tiffany Company in New York had somehow gotten hold of the marketing rights to the class of gemstone that my Eskimo friend was showing me. I told him the stone was probably worth a lot of money and that I would buy it from him if he was interested in selling.

He said he was not, so we parted ways a short while later when I saw one of my correspondent buddies off in a corner of the bar. "I've enjoyed talking with you," I told him. "I hope we can meet up again sometime. You have me wanting to try a piece of muktuk."

"Come on up to Point Hope, man. If I can survive this shithole business for another six months, I'll be heading straight back up. I gotta give this stone to my girl Mary."

When I joined my correspondent friend Jeb in the corner of the bar, he was straight up with me. He told me that if I really wanted to see some action, I needed to hop a supply flight up to Da Nang or Hue near the DMZ, then hook a ride on a supply chopper to the combat base at Khe Sanh.

"Get to know some grunts and have them take you out on a 'Lurp' patrol. Get the commanding officer's permission to go along by taking his picture and promising to get it published back in his hometown paper."

"You better make sure you really want to do this. You can get your ass blown off up there, and you won't even know what hit you." After another swallow of beer, he said, "I mean it about losing your ass. You might wake up in a hospital with half your body gone, and the rest barely functional. Are you ready for that?"

"I can handle it," I said, as if I had any idea what "it" was, "...but what the hell is a Lurp patrol?"

He looked askance for a moment then decided he would give me a straight answer, rather than some smart-ass response to what I learned, later, was a pretty dumb question. So far as most correspondents who had been in the shit were concerned, everybody in the world knows what a Lurp is, but I did not.

"Lurp is short, basically, for Long Range Reconnaissance Patrol," Jeb told me. "Lurps can be as large as a platoon of maybe fifty guys led by a lieutenant or captain, or as small as a squad of seven guys led by a sergeant, or even the squad leader; an enlisted man of lower rank like corporal. It depends on the mission—the distance, the objective, the need to bring along technical specialists and their equipment, the likelihood of enemy contact—stuff like that determines the manpower and firepower on a Lurp.

"Understand that a Lurp is the essence of infantry in war. If you want to understand war on the ground, you have to go on a Lurp; especially one that has contact with the enemy. Maybe you will be lucky and not get hurt and maybe you won't do something really stupid that gets some other dumb grunt on your patrol hurt or killed.

***

It is about 450 crow-fly miles north from the main Tan Son Nhut Air Base in Saigon to the port and supply base at Da Nang, but our "Herc" supply ship had to fly a secure route because of increasing numbers of NVA ground units reportedly operating in south Vietnam. The several abrupt changes in direction to avoid ground fire made the two-hour flight closer to three hours.

Once on the ground and shut down, the Herc pilot took me to the ops shack on the runway. From Da Nang it's a hundred-mile flight to the com-bat base near the village of Khe San in the highlands close to the DMZ. The base is about seven miles east of the Laotian border along the east-west highway named Route 9. According to the pilot the Ho Chi Minh Trail suns north and south. It kind of threads itself inside and outside the Lao border.

"You can just about throw a rock to the west at the Lao border and hit the trail," he said. "Westmoreland figures if he can dump enough bombs on the trail, he can disrupt the NVA arms shipments into the south. He has his head up his ass on that, but what else is new?"

In the ops shack I got lucky. The Herc pilot introduced me to the pilot of a supply chopper. After hearing my pitch, the chopper pilot looked at me with a quizzical look, but he said I could ride along to the combat base if I didn't mind sitting on ammo boxes and C rations.

The pilot called the C rations "c-rats. " That is the name the grunts have given them because, if they were to be eaten with any kind of pleasure, you had to think like a rat while eating. It all made sense to me.

The combat base was originally established as a Special Forces airfield in August 1962. The base represented the start of President Kennedy's military reengagement with the north Vietnamese enemy in Indochina following the departure of the French in 1954. Kennedy's immediate purpose was to provide support for President Ngo Dinh Diem and the government of south Vietnam in Saigon. As the NVA began to infiltrate south Vietnam, Kennedy's military adviser, General Maxwell Taylor, recommended that General Paul D. Harkins take overall command of US military operations.

President Diem was a member of a very prominent family in the minority population of Catholics left over from the French occupation that ended with the Battle of Dien Bien Phu. He had spent the summer of 1963 carrying out a program of brutal suppression of the country's majority Buddhists. There was so much hostility growing against Diem and his family that he and his brother were assassinated by his own military on November 2, 1963. Some believe that US officials met with several unhappy officers in Diem's government in late October and had not discouraged the idea of an assassination.

Even earlier, the self-immolation by Buddhist monk Thich Quang Duc at a busy Saigon intersection on June 11, 1963, and the graphic imagery that accompanied the global news coverage after the event, began to convince many in the population and among

US diplomats that Diem had to be removed from office by whatever means necessary.

By the time of the Kennedy assassination on November 22, there were sixteen thousand “advisors” on the ground in Vietnam, including those at Khe San. Rumor has it that Kennedy never planned a full commitment of combat troops to Vietnam, and that he was planning to withdraw all troops as soon as it became politically feasible.

Now, in the late summer 1967, the main pro south Vietnam guy around is General Westmoreland. He was appointed by former Vice President Lyndon B. Johnson after Johnson took over the president’s office. Whatever Kennedy had intended for America’s military commitment to the war, there was very little talk of withdrawal in early 1964.

By the time of the battle in the Ia Drang Valley, our government officials were openly discussing a military buildup in the south to protect the south Vietnamese government from NVA movements south of the DMZ.

After those fierce four days of battle, Westmoreland decided that bases in the area around Khe San would be a critical part of his plan to cut off the transport of war materials from the north to the south to supply the infiltrating NVA armies. He wanted a large contingent of Marines at the Khe San base to begin what he hoped to be the destruction of the Ho Chi Minh Trail.

The head of the Marines in Vietnam at this writing is one General Lewis Walt. According to military scuttlebutt, Westmoreland and Walt had a major head-butting contest in the Pentagon over Westmoreland’s strategy of committing massive forces to the Ho Chi Minh Trail near the DMZ. Walt was convinced that even if the trail could be severed, or NVA movements of armaments and troops slowed significantly, it would have little effect on the gathering strength of the NVA already in the south.

After all, Walt may well have argued, the Ia Drang Valley is already two hundred miles south of the DMZ, and the NVA was capable, even two years earlier, of fielding and supporting an army that could stop a full-scale US military operation using well-trained troops and advanced tactics and equipment.

Whatever the truth of it, Westmoreland, with Johnson's help won the argument over strategy, and Khe San, supplied out of Da Nang and staffed by Marines, has become a major focus of the US military buildup.

As the Marines moved in, the original occupants of the combat base, the army "advisors" known as Green Berets moved out to Lang Vei, about halfway to the Lao border. There, they continue to work with local Montagnard peasants (the "Nards") on training in US military tactics and reconnaissance methods. The Nards profess themselves to be politically aligned with the US interest in defeating the lowland north Vietnamese. They promise to help both US forces and the south Vietnamese army, the ARVN, in disrupting the movement of military supplies over the Ho Chi Minh Trail.

In their assurances of loyalty, though, the highland Nards may have understated their historic conflict with the lowland Vietnamese, such as those who lived and governed in Saigon, a conflict that has continued at least since the Vietnamese people established their independence from China more than a thousand years before.

***

Shortly after leaving Da Nang, the chopper pilot's voice came over my headset. "You better hunker down and hang on back there. The NVA are sending up some," he paused, then, with a heavy French accent, said, "fleurs du mal. I'm going to have to jerk the airplane around a little bit in order to stay out of trouble."

The pilot began to do some gut-wrenching dips, turns, and stalls with an airplane never designed for such things. It crossed my mind that the pilot might be secretly French, and an enthusiast for Baudelaire. Two loud bangs on the tail of the chopper followed by some ugly vibrations broke my escapist reverie on the ethnic origins and literary preferences of chopper pilots.

The rattling and vibration got more intense, and the chopper started slewing from side to side. The pilot's urgent voice came over the headset. "They clipped us. My steering is a little fucked up, and I need to get her down on the deck. Check around your area and make sure you can see a way out of either side of the aircraft in case I fuck up and we go in hot. If there are any loose boxes back there that could fly loose and hurt some-body, throw 'em out now. We are about ten clicks out from the LZ at Khe San. I'll try to get somebody to help us out if the reception on the ground is too warm."

We skidded around the sky for a few more minutes. I clung desperately to my seat belt, and to the side of the chopper, praying we wouldn't take any more hits.

The pilot came on again. His voice sounded slow and gravelly this time, which led me to think he was wounded.

"Khe San is straight ahead," he said. "I can almost see the landing zone. You need to look around the horizon now to check for landmarks so you can walk out of here. My arm is fucked up, and I'm losing feeling in it. And I am getting dizzy. I'll put her down in the next clearing I see while I can still control our descent."

We flew on for another thirty seconds, then, suddenly, the chopper flipped ninety degrees to the horizon and started to drop down, nose first. I braced, but I figured that this was pretty much the end of it. At the last second the chopper flattened out, the pilot killed the engine, and we hit the ground hard. The landing struts collapsed. There were trees all around us, and the blades above us were still turning and slapping against some of the lower branches.

The pilot slumped forward against his straps. He didn't move.

I couldn't feel any broken bones, but I was sore as hell with bruises where I got banged into the chopper's metalwork and where one of the flying boxes clipped me. I unstrapped and crawled forward over the disarray behind the pilot's seat. A sharp pain shot through my left leg, and I let out a sharp yell. I knew my leg was not broken, but a corner of a box poked into it before flying out the door. I didn't notice until I put weight on it.

In response to my yell, the pilot seemed to rouse himself enough to point toward the first aid kit on the bulkhead. I crawled over, gritting my teeth and favoring my sore leg, then tried to find his wound. His lower right arm was a bloody mess; he couldn't move it, and his flight jacket was shredded. His face looked pale. I have no experience with first aid beyond Boy Scouts and summer camp, but I figured I needed to stop his bleeding before I did anything else.

So, I grabbed a large roll of bandage out of the kit and wrapped it several times around the pilot's upper arm, sort of like a tourniquet. By stopping the bleeding, I figured I would have time to try and help him get out of the chopper and onto the ground before he passed out again. Then I could try to find out how badly he was wounded, and what more there was that I needed to do to help him.

I didn't see any smoke or smell any fuel, so chances were good that there wouldn't be a fire. That was a major piece of luck, but I figured I needed to get moving right now.

Once I got him out and laying on some camo material on the ground, I checked his arm. The bleeding seemed to have slowed a lot, so I loosened the tourniquet and reset it more carefully in the hopes that I could quickly figure out if the bleeding was arterial.

I needed to decide what the hell I should do next. Jesus Christ, I thought. Two nights ago, I was sitting in a warm and friendly bar having a beer and conversation. Now I am in the middle of the goddamn jungle in hostile territory trying to help a wounded soldier whose pure flying skill just saved my life. Now it was my job to save his. I didn't agree to this when I signed on to come to Vietnam in the middle of a goddamn war.

I asked the pilot how he was doing. He spoke not much above a whisper.

"Except for the pain in my arm, I'm okay for now, but there may be hostile NVA looking for us. There should be an M2 Carbine between the seats in the cockpit, and some of these ammo boxes should have ammo for it. Check around the cockpit for a .45-caliber sidearm. Get the carbine, find and load the ammo, and I'll teach you something about fire-fights and perimeter defense. I have my own sidearm, but I only have a couple of spare ammo clips. Check the boxes for .45-cal clips and bring me all that you can find. Call me Jack, by the way."

"I'm Aleck Morris. From upper Minnesota. Nice to meet you." I sounded stupid. I even reached for his right hand to shake before checking myself. He smiled and reached up with his left. We shook.

Wow, I thought. This guy has some balls. I crawled into the back of the Huey to see what kind of ammo and other weaponry I could find. We lost a lot as Jack went through the gyrations to dodge ground fire, but there were two boxes of carbine ammo and a box of .45s. I also found the M2 and started moving the stuff over to where Jack could check it out and try to load the weapons while I continued to reconnoiter our situation and supplies.

Jack whispered hoarsely to me, "We could be here for a day or two. Check the chopper for all the C-rats you can find and bring that aid kit over. See if there is any water. If not, try to find something to catch rainwater. I'm looking around for a place to hole up away from the chopper in case we start to get visitors."

I asked Jack if the radio in the chopper worked.

"Maybe, but it probably won't take our NVA friends very long to find us if we send a radio signal. We need to reserve that asset until we know for sure that it is our last resort if we are to keep from getting killed or taken captive."

What he said took my breath away. I had not even thought about being killed or captured. "Jesus Christ!" I said to myself again. This is real. This is as real as it gets.

Jack continued on. "I see what looks like a small rise about a hundred paces from here. If there is some kind of a hollow behind it, we need to move our stuff and ourselves over there. Take the .45, make sure it has a full clip, grab some spare clips then go check it out before you do any-thing else."

As I looked at the rise and considered the possibility of a hollow behind it, I started feeling hopeful that we could at least hold off a few of the enemy if they decided to attack. We could probably rest here long enough to regain the strength we would need to start moving toward our own forces around Khe San. With luck, the NVA wouldn't find us, or they would ignore us long enough for Jack and I to make our getaway.

When I got back Jack told me to start hauling the ammo and food over to our new hiding place.

"I'll stay by the chopper to cover you if any bad guys show up," he whis-pered. "Get moving. And don't lead the enemy to us by breaking a bunch of branches while you're moving this stuff. Step carefully and be thinking about how to either hide your tracks, or how to trick them into looking into places where we aren't. Once we move in over there, we won't be coming back to the chopper."

It would soon be dark, so it was time to stop talking and start moving our stuff.

***

It took a couple of hours to move everything, to help Jack move to our hiding spot, and to do what I could in the darkening twilight to wipe out any sign of our movements. Jack was going in and out of consciousness, and I realized that I was dead tired as well.

Later, I woke up to total darkness, but some stars were shining through the trees. Jack was still either unconscious or sleeping. His chest was rising and falling normally. I had the feeling somebody was watching me. As I looked around, I hear a soft whisper very close to me. I laid my rifle down next to me before falling asleep, but it was no longer there.

"I moved your rifle so you wouldn't accidently shoot me," the whisperer said. "I'll give it back as soon as I'm sure you won't do anything stupid like shouting or thrashing around or trying to shoot me. Do I have your word that you won't do anything stupid?"

"Who the fuck are you?" I whispered back.

"My name is Sergeant Mike Fredrickson from Brooklyn. Go Mets, Fuck the Yankees. I'm in this shithole of a place as a Green Beret. I saw your bird come down earlier today, but I didn't want to let you know until I had a better feel for whether or not Charlie was in the immediate neighborhood, and whether or not you guys knew how to get your shit squared away."

"So, what did you find out?"

"So far, so good, but it pays to never let your guard down in this part of the world. You're lucky your pilot got his bird as far as he could south and east of Khe San. Most of the enemy is west of here, more toward the border with Laos. The guys who shot up your bird, though, probably saw you go down without a fire. They are probably headed this way to see if there are any goodies they can use."

"Okay. I believe you." I sat up. "Can I have my rifle back? I'll try not to do anything stupid, but I have to warn you that I'm a journalist, not a soldier."

I still could not see the whisperer, but I now felt the returned rifle next to my hand.

“Yeah. I know that. You move like a pissed-off elephant at a rock concert. If I am going to get you and your buddy back to a safer area, I am going to have to give you some OJT on stealth when moving and staying the fuck out of sight when not. How is the pilot doing?”

“I’m not sure. I’m not a doctor or a medic either. He lost some blood. I put on a tourniquet, but he seemed to be drifting in and out of consciousness before I fell asleep. He seems to be breathing okay now.”

“Okay,” the voice said. “Here’s the deal. I work with one of the locals on these kinds of long-range patrols. I dress like a local and speak the language pretty well. I don’t speak at all when I am in hostile company. My partner is a Hmong tribesman named Fres Thao. We don’t carry noisy weapons on patrol, so I’ve sent him back to our base camp to see if he can round up some more help in case Charlie shows up. He’s a fast runner, and I expect him back with a report before dawn. His report will tell us what our real chances for survival are.”

“How far away is this help?” I asked.

“Probably five clicks to the Khe San combat base as the crow flies. Khe San is that way.” The figure of the sergeant finally came into view. I saw him point a partially masked flashlight toward the ground in a direction that looked to me to be a little to the right of where I remembered the sun setting.

“Since the territory is likely hostile, that means it could take us two days to get there. Think about how to get prepared for a long, fast hike with your wounded pilot,” he said. “I think you will be awake for a while, so I am going to take a nap. If Fres comes back before I wake up, I told him to wake me before announcing himself to you. I think your pilot will survive. He might need some morphine for pain, and some help from us, but he should be able to walk out of here. If you have any more questions keep them to yourself until I wake up.”

Mike seemed to fade away like the Cheshire Cat. Other than his whisper, I still had no idea what Mike looked like: not his facial features, not the color of his hair or how it was cut, not his clothes or native dress, or whether he was wearing a hat, or had any weapons that make noise.

Even though it is the rainy season here, the sky was clear of clouds. After some time, I looked up to see if there were any constellations that I would recognize, that might guide me in a safe direction if the pilot and I had to walk out on our own. Polaris was too far north to be seen through trees from this low angle. I thought of the constellation of Cassiopeia, beautiful mother of the even more beautiful Andromeda, both punished by the ocean god Poseidon for their arrogance. Cassiopeia and her throne were upside down for half the night in the northern sky, across Polaris from the Big Dipper. In the southern sky, I was only familiar with the Belt of Orion and the Southern Cross. Neither was visible through the trees.

***

RAFAELA

"Hey Matty, did I tell you that I'll be headed to Vietnam in late summer?"

"Did you finish your military training?" asked Matty Franchetti, my friend from grade school. "Do you have time to earn some money first? Don't you still have a lot of bills to pay after medical school?"

The Cedar Point season had barely begun, but the day was hot. We were sunning ourselves on the amusement park's beach.

I had recently completed medical school with an emphasis in emergency medicine, but I had not yet signed up for a residency. Since high school, Matty had worked as a secretary and executive assistant at the New Departure ball bearing manufacturing plant south of town. Unlike me, Matty had not gone to college after high school. She had taken a series of business courses at a local extension of Ohio State University, and she had done well in them. Her quick wit, native intelligence, her ability to work hard and learn new business and factory floor procedures had earned her a lot of praise from a succession of managers.

After high school, we had lost contact with each other while I moved to the east side of San Francisco Bay to attend the University of California at Berkeley. When I was accepted into the medical school at Case Western Reserve University in Cleveland, we reconnected.

While I have a flair for travel and adventure, Matty is an admitted homebody. Despite these differences we found few others in northern Ohio, male or female, that offered the same kind of human and personal understanding—essentially that we could stand being around for more than a few minutes at a time.

"Hey, Rafe. Do you want to go over to Put-in-Bay later?" Matty asked.

I had been looking beyond Lake Erie toward the distant horizon. "You know, Matt, there is a lot going on this summer out in the Bay Area around San Francisco. Some of it has to do with protests against Johnson's escalation of the war, but there is a lot of fun stuff, too. I need some of that kind of mindless fun before I leave for Vietnam. Medical school was hard work, and my medical specialty in emergency medicine probably means I'll have a lifetime of it."

I adjusted the tie of my bathing suit and then turned to Matty. "I'm not even thinking about Vietnam yet. I hear enough about it on the news every day. I'm still wondering if I should have volunteered to go. Then

I think about my younger cousin. Rafael is in a Special Forces unit. The family and I worry because we don't hear from him, and the army won't give us any details, only that he is on an extended confidential mission."

"I get that," Matty said. "So, let's not talk about it. What's going on in San Francisco?"

I smiled at the clear war within her between trying to make me feel better and hoping that she didn't have to leave Ohio. "A lot of people are coming into the city as war protesters, and there are already a lot of hippies there. The radical fashions that you see around the Haight-Ashbury are exciting all by themselves. The more I think about it, the more I think I want to go back there for at least a few weeks. I just want to see it, because it is so radical, so offensive, and so out of sight that I can't imagine that it will last very long.

"It wouldn't even have come into being at all, except that it is San Fran-cisco, one of the most beautiful cities on the planet. Want to go out there with me? You know, to kind of... check it out? Don't you have a bunch of paid time off saved up?"

"I could probably get the time off, Rafe, but whatever the US is doing in Vietnam has caused the demand for ball bearings to go up fast. My boss has got me reconnecting with our suppliers every day to make sure our critical orders for materials stay on schedule. He also has me working almost full time with the guys involved with the expansion of our pro-duction facilities. I'll think about what you are asking, though. It might be fun, and God knows I could use a break."

"Good." I rolled onto my belly and untied the bikini top. "Put some sun-tan lotion on my back, Matt, please."

"You're such an animal," she said.

"After you've seen what I've seen in med school, you'll realize that you have to see bodies as blobs of flesh. Nothing special, and nothing to get too excited about if the main job is saving limbs or lives or gray matter."

"God, Rafe. You make it all sound so mechanical. Is it really as simple as reconnecting all the wires, and flipping a switch?"

"No. It's not. It's complicated, but maybe that's the real reason why I want to go back to the Bay. They tell me sex out there is getting exciting again. The idea of rubbing and touching naked bodies without having to worry about sterilizing the instruments appeals to me."

Now I was speaking Matty's language. "Okay," she said. "You can count me in. I'll work it out with the boss. How do you want to get there?"

***

A few weeks later, Matty and I were at a dinner party on a friend's farm near Vickery, Ohio, twelve miles southwest of Sandusky. In the interest of saving money for more important things than transportation we had bought a used, fairly ratty '49 Ford sedan for our trip. We didn't improve the external looks of it much, but with Matty's help, I overhauled the engine, increased the displacement, added two four-barrel carburetors, a racing camshaft, and muffler bypasses on both tailpipes. In this way, we hoped to occasionally make the trip more interesting, especially when dealing with people we weren't interested in spending time with.

Our friends Randall and Georgine Keegan had inherited their farm from Randall's parents. As with most farms in this part of Ohio, they raised milk cows, hogs, chickens, and a few dogs and cats. They had about three hundred acres of corn, wheat, and soybeans in cultivation, including pasture for the cows. They weren't probably going to get rich, but they worked hard, and they made a comfortable living. Georgine was pregnant with their first child.

"Why in the world do you want to drive all the way across the country to San Francisco?" she asked. "I thought it was bad enough that you were away from us for four years, let alone in that pit of sin and corruption."

"It's not such a bad place, Georgine. I learned a lot at the university. I met some good people, and I learned so many important things that I was never going to know if I had stayed here. Besides, you know I was getting crazy bored in high school. That's why I kept getting in trouble, and that's where I learned that the good news about high school detention rooms was that I could read any book I wanted. Besides, people in San Francisco are as fine as any people I have met anywhere. There are a lot of churches and places of worship. Not all of them are Catholic or Lutheran, or any denomination you might have ever heard about in Vickery. Come out sometime and let me show you around."

Georgine looked down at her feet. Randall looked out the window as if checking the weather, though the warm, sunny spring weather had not changed in a week. Matty's look told me that I had probably said something offensive. Even so, I had also wondered at the sharpness of Georgine's remark.

Matty reopened the conversation by asking about my upcoming military service. Randall suddenly looked toward me; his steady gaze full of questions.

"What's that all about? "he asked.

"I'd been thinking about signing up," I said, "so I finally did a few months ago. You know, my cousin Rafael joined up with the Green Berets after President Kennedy set up those special force units. Rafael and I have always stayed in touch, and I have always been impressed with his commitment and his belief that he is doing something important. Now that I am a medical doctor with some valuable skills, especially in wartime, I think it may be time I offered to do something important, too."

Randall was tightlipped. "I don't know if what is going on in Vietnam can be considered a war yet, though I guess LBJ wants to make it in one as soon as he can. Reminds me a whole lot of the 'police action' in Korea. What a goddamn joke. Hell, we are doing a worse job at defending 'American interests' than the French did at defending the right to eat croissants at breakfast in that shitty little country." He got up. "Excuse me, I have to go out and slop the pigs."

“Wait a minute,” Georgine said. “I’ll come with you.”

After they stepped out the door, Matty said, “Wow! Does any of that change your mind about signing up?”

I worried my lip. “No. But maybe it is time for us to go.”

As we walked toward the car, Randall called out to us: “Wait!”

He and Georgine both approached. “Look,” he said, “I’m sorry for being rude.”

“I am, too,” Georgine said. “You are both among our very best friends, and we have no right to judge what you are going to do. We want to al-ways support you in whatever you do, even if we don’t have a clear idea of why you are doing it.”

I was relieved, and we joined the couple in a hug.

Randall took a step back and ran his hand through his hair a couple of times. “I don’t know about this kind of war. I’m a farmer so I am always exempt from the draft and military service, but this kind of war seems different somehow from the war I grew up with in the early forties. We are not sending our tanks and planes against our enemy’s tanks and planes. We are sending all of the killing machinery and might of our military against a people who don’t even have their own uniforms, and for what reason? Are we making the world safe for democracy by bombing mud huts? Is democracy and ‘We the People’ no longer a consideration in these kinds of wars?”

He paused for a moment. “I fear that this war will end very badly because of that terrible premise. I hate to see my very best friends getting involved, but I also understand that because of the draft, many good young men are already involved no matter what they think.” He looked at me. “And good young women like you are going even without a draft. Hopefully, you can balance the lunacy and the lobbyists for the armaments industries that I see driving all this war talk from Washington. You take care of yourself, Rafaela.”

I will. Thank you."

"Me too," Matty said, "though I'll be back in the neighborhood here after Rafe and I check out some of that sin and corruption that Frisco is famous for."

Georgine burst out laughing. "I love you both," she said.

Randall didn't laugh, but he gave each of us a separate hug. "Take care," he said.

***

In early June, the sun was just coming up behind us as we drove toward the Rocky Mountains. We were somewhere in the Utah desert west of Salt Lake City, driving fast across a long, flat, dry lakebed. I had bypassed the mufflers, and my speedometer read a little over 100 miles per hour. We had picked up a hitchhiker at the University of Chicago to help with the driving. He had long, scraggly hair, a tie-dyed T-shirt, and ratty old shorts and sandals. He also carried a large jar of multicolored pills. He was asleep behind me. Matty was riding shotgun, also asleep.

I gloried in the freedom—and the sound, through my open window—of the screaming engine and open pipes echoing off the surrounding hills. Suddenly, I noticed the blip-blip-blip of a flashing red light in the rear-view mirror. I knew our Ford was fast, but it was about fifty miles to the Nevada border. I was briefly tempted to make a run for it, but I probably couldn't outrun the patrol car before my souped-up engine blew up. Worse, if I was caught doing something like that it would mean the end of my careful plans to go looking for Rafael.

I pulled over, as did the police officer. The officer approached our car. He was calm but frowning. "Did you see the twenty-five-mile speed limit in the town you just blew through?"

"I'm sorry, officer. I saw a couple of buildings, but I don't remember seeing a town. I know I am over the speed limit, but I'm a medical doc-tor, and I am trying to get to Fort Ord for some orientation to my duty station in Vietnam. I'm running late, and the road ahead looked clear and straight, and there were no cars coming." Whatever little fib it takes, I thought to myself.

The officer looked skeptical. "Do you have some papers that will confirm any of that?"

"Yes. I have a letter of congratulations from my medical school. I also have a letter from the army telling me where to go when I have my personal business wrapped up."

"Let me see 'em," he said. "By the way, who's your friend in the back?" He pointed at the scraggly form in the back seat, the bottle of pills hugged tightly to his body.

"A hitchhiker we picked up in Chicago to help with driving and gas money. He told us the pills are diet supplements."

After looking at the papers for a moment the officer said, "Look. I am going to do you a favor by letting you go with a warning, but please do me a favor by reconnecting your mufflers and by trying to stay somewhere close to the speed limit while you are in Utah."

I felt a flood of relief. I had gotten more stressed by the situation than I thought I would. "I will, officer. Thank you very much." The officer stood back from the car. I flipped the lever that reconnected the mufflers and pulled back onto the main highway.

Matty lifted an eyelid. "Hey, Rafe. I'm really impressed."

"Cool," said the fellow in the back seat.

The remainder of the trip across the Rockies and the Sierra Nevada was routine, with the exception of a wreck on one of the mountain passes west of Lake Tahoe. A car had gone through a railing and rolled down a hill. An ambulance and a state trooper were parked next to the break in the railing.

Matty was driving. "Pull up to that state trooper," I said. I rolled my window down and spoke to the officer. "I'm just out of medical school, sir, trained for emergency medicine. Do you need help?"

"Yeah, maybe so, but it looks pretty bad. Hold on for a second." The officer turned to his walkie-talkie. "I've got a young woman up here who says she is a med school graduate. Can she do anything useful down there?"

The connection had a lot of static, but a voice came through. "I don't think so. His neck is at a bad angle, and I can't get any vital signs. I don't think there is much to do but cleanup and paperwork."

"I'd like to go down anyway, officer."

He looked closely at me. "Knock yourself out," he finally said.

But there wasn't much anybody could do. I could see at a glance that nothing could be done to bring the twisted form back to life.

The ambulance crew thanked me for coming down the hill, and for being willing to stay in contact.

The three of us drove on.

====

ALECK

"Get up. We have to move... NOW!"

I rolled over to grab my helmet and rifle. Somebody grabbed my shirt and pulled me up. It was barely light enough to see, so I was having trouble keeping my feet under me and moving in the right direction. A man I assumed was Mike was crouched and running ahead of me. "Where's Jack?" I whispered.

"He's ahead of us. Don't worry, we have him," Mike said. "Keep quiet. Charlie's coming."

We kept crouching, ducking, and trying to move as fast as we could through the mud and brush. The rising sun was behind us, so I knew we were running in the general direction of Khe San, but I had no idea how far we had to go.

After about a half hour of running, my feet began to drag. Mike noticed and stepped off the trail where the brush was partially cleared. He pulled me into some denser undergrowth.

"How you doin', cowboy?" he asked, his voice low.

I wanted to throw my weapon and extra clips away because they were heavy and cutting into my skin, but I knew I couldn't. "I'm out of breath and feeling totally beat up. How much farther?"

"My guide cut some brush for us, so as long as we can stay on this trail, we should be able to get to Khe San by nightfall. Otherwise we play Boy Scout for another night. The only trouble is if Charlie finds this trail, he can move along it faster than we can."

"Do you know if Charlie is close to us?" I asked.

"No, I don't, but if you stay quiet, I'll listen for noise that might tell us something."

I sat back to catch my breath. Suddenly it occurred to me why Mike was so hard to see in the dark. He was black. I had to ask: "I thought Green Berets were all white boys from the Ivy League." I tried to keep my tone light, but I don't think Mike took it that way.

"Shut your mouth, white boy. I'll talk to you in a minute about that."

I tried listening to the ambient sounds to see if I could detect any signs of trouble. There were many birds, probably some monkeys chattering and snakes slithering, but no human sounds that I could hear. Nor were there the kinds of sounds that go with human activity: metallic clicks, bumps, scratching through cloth, branches breaking. Mike looked at me with his finger to his lips, so I stayed quiet.

He put his hand on my rifle, then waved a finger at me to indicate that I should stay calm. Suddenly, there was a disturbance in the brush a couple of yards away, and I almost crapped my pants. The leaves next to Mike opened slightly, and a local man stuck his head out. By now Mike had a death grip on my rifle. I couldn't have moved it if I wanted to. Mike and the visitor talked quietly in what sounded like the visitor's native language for a few moments, then the visitor moved back into the brush and disappeared.

"That was Fres," Mike explained. "We almost have your pilot back to the Khe San base. He can move on his own, but he is pretty weak. Mostly we had to carry him on a stretcher. You and Fres and I are on our own, but your pilot has a squad of Marines to take care of stretcher-bearing and perimeter defense."

"Where has Fres gone?" I asked. I wanted to know about our own chances for survival.

"Fres was with the other group until he felt they were safe. Now, he is moving back along the trail to see if Charlie is anywhere around. If he finds Charlie, he may take him and his friends out. We won't even know about it. I think we are safe for now, so let's move out as fast as we can. Stay low. How's your leg?"

"Okay," I replied, "but I don't think I can run very far, or very fast if we get into trouble."

"I'll keep an eye on you, but I may get ahead of you and you won't know where I am. Try to keep calm because you need to keep moving. In general, I'll be moving toward the setting sun. As long as you move in that same direction, Fres or I will find you. If you lose sight of the sun or start feeling disoriented and panicky, stop where you are, make yourself invisible, and wait. If it gets dark, get comfortable, and try to get some sleep."

Jesus Christ, I thought to myself. What the hell was I thinking when I asked for this assignment?

I did pretty well, but it got dark, and I had lost track of Mike a couple of hours before. I stopped in a place where I thought I could survive the night and covered myself with some branches and leaves. I fell asleep immediately.

The sky to the east was growing light when I felt a hand on my shoulder. I opened one eye and saw Mike. He motioned me to my feet, and then to follow him. After a few hours we slipped through the perimeter defenses and into the compound set up around the Khe San Combat base.

We are home, I said to myself, at least for the time being, and as much as a forward combat base can be thought of as a home.

***

RAFAELA

Matty, our tie-dyed passenger and I drove over the crest of the Coastal Range on Route 80 about ten miles above Vallejo. A local Latino rock station was on. As we continued south toward my old stomping grounds, the Golden Gate bridge came into view in the west.

"It's beautiful," Matty said.

I felt more at home here than any place else I had ever been on the planet. "Especially so in this beautiful morning sun," I said.

"This is really cool," said the guy from the back.

"You know, Tie-Dye," I looked at him in the rearview, "we don't even know your name.".

"My friends call me Habit Forming, but you all can call me Hab." Matty turned to ask, "Okay, but what's your real name, Hab?"

"Shit. I'm busted," he said. "My real name is George Steiner. What's yours?"

Matt turned to George and extended her hand. "I'm glad to meet you, George. My name is Matty, but you can call me Matt. Her name is Rafaela. You can call her Rafe."

"Are you guys new to the Bay Area?"

"I am," Matty said. "Rafe graduated from Berkeley."

"Cool," said George. "I was born and raised in the city, over on Van Ness. My parents taught law at San Francisco State. I was in Chicago getting a master's degree in math, but mostly trying to beat the draft. That only works for so long. I finally got my notice. I have to report for basic at Fort Ord in about a week."

"I know the feeling," I said. "I have to report to the Letterman hospital on the Presidio in two weeks. Do you want us to drop you at your parents' house?"

"No. My parents don't live here anymore. The gang I run with now are all over on North Beach. You can drop me off anyplace, but if you want to go over to North Beach, I can probably get us something to eat and tell you some fun places to go this evening. Why are you reporting to Letterman?"

"I just finished med school, but I haven't done a residency yet, so I have worked out a deal where I go as a nurse and basically pick my assignment. The army said I should complete the residency and get certified as an MD, then come back. The problem with that is that there are almost no female medical doctors anywhere in Vietnam, certainly not anywhere near any front lines, and I want to get to Vietnam soon. And I want to get to the bush. I have a cousin there that I haven't heard from in months. The army only tells me that he is on an extended mission that is classified."

"Whoa," George said. "Where do you want to go?"

"The highlands around the DMZ. There's an evacuation hospital near Phu Bai that is the first stop for those wounded in the highlands. The last I heard my cousin was assigned to a unit near the Laotian border."

"Man. You got some set of balls, lady. I'm going to spend my time in this man's army trying to avoid doing anything dangerous, or even a little risky, or difficult in any way, and here you are jumping into the middle of the shit. Jesus Christ."

"Let me change the subject," Matty said. "What are some good places to go in North Beach? I've heard of City Lights, Fisherman's Wharf, Visuvio... are there other places?"

"There are many places. No matter what kind of entertainment you like, North Beach is the place to go. You can find it all there, especially when you have a name like Rafaela. North Beach is basically an Italian fishing community. There are a lot of retired fishermen from the old country scattered around the neighborhoods."

"I'm going to drive over there," Matty said. "Can you show us some of the sights, George? Maybe introduce us to some people?"

George got us settled into a cheap room in a fishermen's retirement hotel. In the hallway on our floor, we were greeted by a couple of old fishermen who were cooking a pot of fish stew. They welcomed us to San Francisco and asked if we were hungry. The stew smelled delicious. "Starving," we said in unison.

After eating and helping the old men clean up, we caught up with George, his shirt off, down on docks in the marina sunning himself. We walked around for a while then decided to take a cable car downtown for some sightseeing. I could feel myself getting bored, though. My mood did not improve on meeting some of George's friends at a lunch counter down on Market Street. Some of them were stoned, and others had been in that condition not long before. Some looked as though they might have been stoned for the past several weeks.

We hung out for a while at a bar and strip club that evening, but I said I was going to beg off from a late evening. After several days on the road, I felt like I needed to catch up on some sleep.

Matty had been somewhat wide-eyed as she took in all the sights, sounds, and colorful inhabitants of the city. At the sight of a black man walking arm in arm with a very pregnant white woman, Matty's mouth fell open. Further along, at the sight of two Chinese men in vivid drag and high heels, she almost swooned.

There is nothing like this in Sandusky," she said, sotto voce. I smiled to myself. "No, there is not."

Matty had also had a full day of it, then, and decided to come with me back to our room. "I'm going to stick with Rafe," she told George. "Thank you for showing us around and introducing us to your friends."

George had been getting more stoned as the night progressed. He gave us a brief two-finger wave and a lopsided smile. He turned back to his friends, who were equally out of it.

As we walked out the door, I noticed a poster for the Monterey Pop Festival for the coming weekend, about one hundred miles down the coast.

I pointed at the poster. "How about this pop festival? We will have to rearrange our schedule, but it might be worth it."

"I'm up for it. I think I can work out a few more days on the West Coast," Matty said.

"Good. Say, I know a great Chinese place a few blocks from here. Let's go there and eat, and figure what to do. I know one thing I need to do: after I visit the Presidio; I'd like to deliver our car to a friend on the faculty at Berkeley. I'm sure he can be persuaded to work with your schedule, and to drive you to the airport when it's actually time to leave."

Matty crinkled her nose and stared at me. "Is this a 'special friend,' Rafe?"

"He's pretty good in the sack if that's what you mean."

***

ALECK

It's fall in the highlands, and it rains a lot.

I've taken a few days to recoup and try to catch up with my editors after the trip up from Saigon and the helicopter crash back there in the boonies. Jack made it here, but they have medevacked him out to the mobile hospital at Pleiku. They tell me he is okay, but probably done with this war. I got to wish him well as they loaded him into the chopper. He told me not to take any more helicopter rides.

“You never know how those things are gonna go,” he said with a big smile. We laughed and shook left hands.

For a time, I shared a hooch with a lieutenant grunt named Parker. He’s from Georgia, and his nickname is Chip. He is a combat commander responsible for Company C, 3rd Regiment, 26th Marines, and he is telling me about a guy named Rafael. Rafael was born and raised in north central Ohio. His parents are part of a big Italian family in that area, most of whom work in some way with Great Lakes Shipping.

“I tell you this for background on this guy,” Chip said. “Most of these Green Beret grunts are college kids who are looking for a little more action than you can get in Kennedy’s Peace Corps, but Rafael is truly a ghost among ghosts out here. Back in the world, though, he is a solid guy with solid roots in the Midwest. When you go out on this Lurp you’ve talked yourself into, I want you to keep an eye out for him.

“Now that I think about it, ‘keep an eye out’ is probably a misstatement. You won’t see Rafael until he is on top of you. I could say ‘keep an ear out,’ but actually, that probably won’t work either. You won’t see him or hear him, but you might smell him. He likes to wear a flower in whatever native hat he decides to wear.

“You’ll also know if you feel his touch, but you probably won’t like the feeling. If he doesn’t know who you are, or why you are on his turf, he may put you in a death grip before he asks any questions.” Chip patted me on the shoulder. “But I am sure you won’t have any problems. Get some sleep.”

Then he said, "One last thing. It's not like I know anything about his mission, but he saved my ass once, and I got to know him while we were trying get out from behind enemy lines. I don't want to lose track of him if there is anything I can do to avoid it. If he is in trouble and needs my help, I want to make sure he gets whatever he needs."

Parker was right about my having talked myself into going on a Lurp. The combat base commander—a light colonel known to the Marines under his command, but never to his face, as Butthead Bukowski—said I should get some sleep because he was not sending me anywhere until he felt my chances of screwing up the mission were zero, and the chances of my surviving a crippling injury were better than fifty-fifty, whatever that meant.

"All I need is for some dog-shit reporter from some chickenshit anti-war rag in the Midwest to screw up my mission, and get killed, or captured, or physically fucked-up on my watch," is what he actually said. "If the 'request' that I try to accommodate you hadn't come down from fairly high up, you'd be sittin' around here with your thumb up your ass until it was time for you to go home." He paused and gave me his best penetrating glare. "Got it?"

"Yes, sir, I got it. Thank you for setting me straight, sir," I said. "Do you still want me to get your picture published in your paper back home?"

"Get the fuck out of here, shithead," Colonel Butthead Bukowski said.

As it turned out, though, I sat around for several weeks with my thumb up my ass anyway. I also spent time hanging out with Lt. Parker, who I consider to be a pretty straight shooter. He gave me a lot of good back-ground on what a combat base is and what a firebase does, and on the strategic importance of the Khe San combat base at this stage of the war.

There had been a lot of action at Khe San before I got here, but it was fairly quiet now that we were into the rainy season. The lull gave Parker and me a lot of time to talk in between my attempts to send some copy off to Penny Blue, over a flaky and irregular connection, about the situation here. We also smoked the occasional doobie and toked on some opium whenever somebody brought some onto the base that they were also, rarely, willing to share.

Parker also gave me some basic training with the weaponry used on a Lurp. I spent a lot of time perfecting my aim with an M-16, blowing up bottles and stumps with hand grenades, and getting to know some of the grunts who actually did the ground-pounding.

***

It is now November 1967. Each day is generally dark and rainy. Chip and I have cracked some C-rats and are chowing down. The USAF B-52 bomber air raids are continuing, weather permitting, along the Ho Chi Minh Trail, and we can hear the bombs exploding. The B-52s can fly above the clouds, so they can still do some damage with radar-guided bombs. Other aircraft, like lighter bombers and our supply choppers, can't fly in this weather. Supplies over Route 9 from Da Nang are also disrupted because of VC sappers working along the route.

This is all part of Westmoreland's strategy.

What a joke. Some of the scuttlebutt has it that the US has dropped as much ordnance on the trail as all of the ordinance dropped in Europe during World War II. Once they figure out how easy it is for the NVA to anticipate those raids, how easy to then move most of the hardware and personnel off the roads and into the woods, and to have the damage to the trail repaired within a few hours, maybe they will stop disturbing our sleep with all the noise.

Our military planners in this war should have paid more attention to the costs and benefits of the bombing campaigns against

Germany in World War II, specifically one raid in particular, the October 14, 1943 raid on the extensive ball bearing works in Schweinfurt. For all of the B-17 bombers, crews, and bombs assembled for the raid by the US 8th Air Force, the production of ball bearings in Schweinfurt was back to pre-raid production levels in a matter of weeks.

Even the atomic bombs on Hiroshima and Nagasaki might have had the same effect on Japan's war making machinery if it wasn't for the pure shock value of a single plane dropping a bomb that could cause so much damage and fiery death to so many people.

The shock value to the enemy after, say, the tenth nuclear weapon dropped on a city and killing hundreds of thousands will likely be zero in terms of an ability to discourage an enemy's desire to make more war.

The thing that killed the Japanese war effort was the immediate sur-render of Emperor Hirohito within a day or two after the US dropped its second, and last available - of the original three mission-ready Atomic bombs produced at Los Alamos - on the city of Nagasaki.

Whatever anybody else believes, so far as I'm concerned, the atomic bomb had precisely no effect on Japan's desire to fight to the death to protect their homeland. In Germany, the war effort finally collapsed when Adolf Hitler committed suicide with his mistress instead of surrendering to the Russian armies closing in on his bunker deep underground in a Berlin neighborhood.

Now, two decades or so after Hiroshima, there is no "pure shock value" associated with the use of atomic weapons. The eyewitnesses to the blasts are either dead or dying, and the narrative about the reasons for using the bombs then was mostly transactional. In other words, the rationale for using them at the time, according to President Truman, was to reduce the loss of American lives that would result if the US invaded Japan with a ground force.

There is no longer a moral or ethical dimension to the decision-making involved with strategic bombing, nuclear bombs or not; there is only the cost-benefit analysis, like being in a grocery store and deciding to buy chocolate or peanut butter. No wonder the grunts on the ground here are suffering such low morale; no wonder there is so much escape from reality. I am beginning to think that there is no reality here, only a faint disturbance in the global propaganda and mutual stroking that sustains these actions and this war effort.

I'm sure that Westmoreland is considering the use of nuclear weapons on the Ho Chi Minh Trail, just to prove to his many critics—in the press, in the government, and within the military itself—that his strategy is correct in dealing with his military problem. It is a problem so severe that Westmorland believed that the use of nuclear weapons is justified. If he did get approval from Washington to use them, I just hope he would do us the courtesy of giving us news guys and the grunts around the trail enough time to get the hell out of the way.

Colonel Butthead finally gave me permission to go out with a Lurp. The mission will involve a squad of Marines from the combat base doing recon along a trail to a Green Beret outpost at Lang Vei. Lang Vei is very close to the Lao border. We leave at 04:00 tomorrow. Butthead told me I would have to gain the trust of each of the members of the squad before I could go with them.

"If any squad member doesn't want you to go, then you don't go. Got it?"

"Yes, sir. Got it, sir," I said. "Where do I find the squad?"

"They are gathering their gear and supplies over next to the ops shack. You better get over there."

He started away, then paused and turned back to me. "One last thing. I have been impressed with the work you have been doing with Lt. Parker. He is also impressed with the improvement in your weapon handling, and in your physical ability to handle the sheer hard work involved on a Lurp. He has been your strongest advocate," Butthead said, "and he has agreed to go along to look out for you. You better get a move on. Find Parker and take him with you to help break the ice with the squad."

Parker caught up to me as I made my way to the ops shack. He had removed his insignia of rank. The squad members were in battle fatigues, and were already packing up their gear, getting ready for the early morning start. Parker introduced me to the squad leader for this mission, Sergeant Jerry Collins. Collins is a big guy, at least six-three or six-four. He shook my hand but didn't say anything. He pointed to the other members of the squad, then gives a nickname as he points to each one. A short, stubby guy is "Hap." A skinny guy with sideburns and barely a beard is "Ralphie." Another big guy, more heavy than tall, is an American Indian named "Sitting Bull, sometimes just Bull."

After the introductions, Collins turned to Parker and asked if he wanted to say anything about me.

Parker told them he has been working with me and has been training me on weapons and fitness. "I'll be going out with you guys to help Aleck get his story about what we do, and to know how to keep out of the way while you do it. He's a good guy, and he has been working hard to be able to do this with you. I do outrank Collins, but he is your commander on this mission. Don't forget it. If I am forced to remind any of you to look to Collins for leadership if things get tricky, I'll remind you, once again, that Collins is the boss, and I will remind you in ways that you won't forget for the rest of your tour of duty.

"Give Morris something useful to do on the Lurp. Otherwise, he'll start thinking this is just a long hike like the ones he used to go on holding hands with his sweetheart back in Minnesota. Give him some gear to haul and be responsible for, and make sure that he gets a weapon with a couple of clips to use if things get bad in the shit. I have been training with him. He knows how to handle your weapons, he understands basic tactics, and he is a better shot than a lot of the other combat grunts I know on this base."

One of the guys, "Mary Jane," asks, "What the fuck you doin' here, man? You don't even have to be here in this shithole, and you all are going out on a combat mission? You're fuckin' nuts, man. We're gonna get our asses shot at, and some swingin' dick's gonna get hurt, guaranteed. "

I didn't have any quick answers for Mary Jane, but I knew I had to say something. "Look, I know you see me as probably more of a liability than an asset to this mission. Going on this patrol is probably the most import-ant thing I will do while I am in Vietnam, and I promise I will do every-thing I can to not let you guys down."

Another guy—name of Choi, a weapons guy—asked what I would do if Parker and Collins got into a pissing contest during a firefight. "Who you gonna go with?"

"Collins," I answered with no hesitation.

Another guy butted in. "Is Parker your babysitter?"

"I don't have a babysitter. Collins is squad leader. Collins is the boss."

From the back, I heard, "Good answer." A guy smiled my way and said, "Not common over here in the shit."

"Let's cut the bull and get you some gear," Collins said. "Pollock, you help him out."

The guy from the back nodded. "Got it, Sarge." "Sounds good to me," I said.

***

RAFAELA

Despite our best efforts, we did not find a decent hotel near the festival in Monterey. Rather than sleep in some fleabag a ways out of town, we decided to find some space on the beach to camp out.

Even the beach was crowded with transient festival goers who could not afford to or did not want to buy hotel space, but who were in the mood for a big party. For two attractive women from northern Ohio, our night was full. We got up late the next day, and spent some time finding some-thing to eat and deciding which acts we were most interested in.

Over a cheesy croissant, Matt suggested that Big Brother and the Holding Company were featuring their incredible singer Janis Joplin, and that they would be onstage early Saturday morning. "The George known as Hab told me that she and the band have been playing the clubs in the Haight, and her singing will literally 'blow your socks off' is how I think he put it. I think if we do nothing else here, we need to hear Janis Joplin sing."

***

Back in San Francisco at the Presidio on Monday, I drove to my appointment with a Captain Dave Margolis, head of the medical unit. He grinned and shook my hand, then invited me to sit down. "Welcome, Doctor."

"I'm not quite a full doctor yet, though I hope to be one as soon as I complete my obligations here, and get into a residency," I said.

"Well," said Margolis, "I hope we can help make that possible. Your papers are squared away, and I've got clearance to put you on a bus to Fort Ord tomorrow. Are you ready to travel?"

"Yes, sir, I am."

"Ask any soldier where it is. You will do about ninety days of training on how to look and act like a lieutenant who is also a military nurse. Once that is done, as I understand it, you'll be flown out to Saigon, then dis-patched out to the 22nd surgical hospital at Phu Bai up by the DMZ. The commanding officer at Phu Bai is Captain Tom Burgess from Tuscaloosa. He graduated from Ol' Miss. He is a good guy, but he talks funny and can be rough around the edges. He has a lot of experience with combat trauma, though.

He has been briefed about you and is looking forward to the medical help you can provide. The nurses up there tell me he knows his stuff, but he can be a real shithead about procedure if you get on his bad side. Unless you have any more questions of me, good luck and God-speed."

***

I had to "fit in" during my time in the Officer Candidate School training company so that I could become one of the army's desperately needed ninety-day wonders--second lieutenants, in other words. In this training company the drill instructor was a frustrated student of medicine herself. She had an undergrad degree in pre-med but then lost a scholarship and could not afford to go on to med school. She joined the army and is taking night courses in nursing. The DI hated the idea that I was already a doctor and was now learning how to be a nurse, and I couldn't say I blamed her.

At the company's first muster, our DI walked into the room. "Attention. My name is Sergeant Genevieve Carson. I am your drill instructor, and believe you me I will be on your ass for the full ninety days of your time here—assuming, or course, that you have enough lead in your ass or grit enough in your backbone to put up with me and stay the full ninety days.

"My job is to find out what you do have to offer us and what you don't have to offer us. I'm damn good at finding things about you that you never thought you had, good or bad. If you don't have what it takes, I am going to send you back to your pathetic, enlisted grunt job so fast you won't know what hit you. You will probably cry on the way out of my office. You might even start begging for another chance. But you know what?" The DI looked around the room, allowing a few moments while her new charges let their imaginations run wild. "I'm not even going to give a shit. Now get in front of your bunk and sound off with your name, rank, and MOS."

When I gave my name, the DI let her eyes linger on me for a moment before moving on.

After names had been given, she walked up to me. "Listen up, all you grunts. This is a DOCTOR training to be a nurse." She paused. "My fuckin' ass."

The DI looked around at the other trainees, then, since she was much shorter, back up at me.

"Who the fuck do you think you are, Barardi?" she asked. "You haven't even been through army basic yet like the rest of these poor bitches and you are already trained and ready to start earning millions of fuckin' bucks for sticking your finger up people's asses. Why don't you drop and give me fifty so you can get a feel for what those asshole male doctors are going to do to you out in some field tent in the boondocks—and I don't mean they are only going to ask you to get them coffee. Get down on your skinny belly and give me fifty. Now!"

I was in good shape, but I hadn't done fifty push-ups since college basketball. Nevertheless, I completed the task and stood up to face the DI, out of breath.

"Come to attention when you finish doing what I tell you to do, DOCTOR. Say, 'Thank you, sergeant.' Say it loud, say it right now."

"THANK YOU, SERGEANT!"

"Good. Don't ever say, 'Thank you, sir.' If you do you will have to do it all over again. Are we clear?"

"YES, SERGEANT! THANK YOU, SERGEANT."

"That's good, Barardi. You're not as dumb as you look. You be my good little poodle dog for ninety days and maybe we will get along." Sergeant Carson moved away to her next object of abuse. All the other women were standing at stiff attention and staring straight ahead.

Sergeant Carson moved away to her next object of abuse. All the other women were standing at stiff attention and staring straight ahead.

I wondered to myself just what I had gotten myself into.

***

ALECK

Collins was pissed. The radio was fucked up, and the radio guy, Easterman, couldn't call in an airstrike on an NVA position set up below them across Highway 9, about a mile west.

The NVA was too far away for our sharpshooter, so Collins determined to get closer. I wasn't too happy with that idea. I guess my recent experience in the deep bush in hostile territory still lingered with me, but I really couldn't say anything about that, and I really couldn't say that I didn't want to go. I started picking up my gear to move out.

Collins spoke up. "Things could get hot, lieutenant. You and Morris may want to stay up here while we take care of business down below. We'll get back to you as soon as we can."

Parker looked in my direction, but I kept my face expressionless. I continued gathering my gear, but a million thoughts were running through my mind. Then Collins looked at me, and I blurted out: "I need to come with you, but I don't want to screw up tactics or get caught in any cross-fire. Is there a place where I can see what's going on, and you guys can find me after it is over?"

Collins looked down for a moment, then to the west. "You see that rocky outcrop about a half click from here? You have to stay behind us, but you can follow us so long as you can keep us in sight. If you lose us, head for the outcrop. We will get there when we get there. If we don't get back to you, figure you are probably on your own. Plan on staying the night, then make your way to Lang Vei at first light tomorrow morning. There is a firebase there with a bunch of Green Berets. If you can make contact with them, they will get you back to Khe San.

"The main thing to remember is that all of Route 9 is contested territory. We might control a piece of it one day; then the VC will take control of the same piece the next day. Typically, what either of us does when we first retake control of a piece of the road is hide and set up an ambush. Before you show yourself anywhere on the road, stop, look, and listen. Don't make a move until you know for sure who is in control.

"All those B-52s that have been bombing the shit out of the trees and dirt along the trail haven't really done any more than stir up a big hornet's nest. NVA troop strength is building in this whole area every day. So, watch your ass, stay in the woods, and stay off anything that looks like a regular trail."

Mary Jane sat nearby with an ear cocked in our direction. "Uncle Ho has figured that Uncle Sammie thinks the war can be won by bombing the trail. The Ho Ho man wants to send enough troops and weaponry into this area to keep Johnson and Westmoreland thinking that the bombing is the only way to stop the NVA providing support for their NVA buddies in the south. If the NVA collapses in the south, why, shit, man, we can be home by Christmas." He Jane broke out into a raucous laugh, then stopped. "What a fuckin' joke."

Mary Jane looked like he had more to say, but Collins cut him off. "Let's go."

Parker grabbed his gear. We waited for a few minutes before following the squad down the hill toward the NVA position. It became clear right away that we could neither keep up nor keep the squad in sight. They disappeared almost immediately. Collins' comments about the NVA buildup in the area also began to work on our minds.

Parker spoke first. "We need to clear our heads about what, exactly, we're doing. We can't be anywhere near that squad if they get into a firefight. If Charlie finds us, we can't survive on our own away from the squad anyway, even with the weapons we have, even if Charlie does something stupid, which he is not probably going to do.

"We need to move now to the outcrop and see what happens from there. Maybe we will get lucky and have a good view of the NVA position so you can get your story."

***

I was familiar with this kind of bushwhacking because of my escape after the helicopter crash on the way up to Khe San with Jack. The number one rule: stay away from anything that looks like a trail, because it is probably booby-trapped. Booby-trapped or not, Charlie will recognize your footprints and will come after you if you give him a trail of shoe prints. Remember that Charlie is probably wearing old, worn-out shoes that may have been repaired by strapping pieces of old tires together. He might track down his own mother if it meant he could get a new pair of shoes.

We were now a couple of hundred yards from the outcrop. Parker signaled for a break, and we sat down. The bush was fairly dense, so we couldn't see much.

Suddenly, an Asian face appeared in front of me with his hands up and his palms out toward me. His face was no more than two feet from mine. I recognized Fres, the Hmong tracker with Mike Fredrickson, the Green Beret who helped get me the rest of the way to Khe San after the crash. Fres had his finger to his lips, so I reached over to touch Parker, who had his back to me. He had not yet seen Fres. I didn't want Parker to suddenly turn around and see an enemy.

Parker was astonished, but Fres and I both had our fingers to our lips and our other hands up with the palm facing him. Parker looked at me quizzically, but he did not pull his service revolver. I deferred to Fres, who indicated that we should stay put and wait. He turned around and disappeared into the brush.

About ten minutes later, the bush parted once again, and Mike Fredrickson appeared. "Hey, guys. What's up?"

"Hey, Mike. It's good to see you again, I think. This is my pal, Lieutenant Chip Parker."

"Hey Chip, how's it hangin'?" Mike didn't wait for an answer. "Here's the deal. You guys have busted into one of our ops. We are trying to take out that NVA installation down there just off the highway. They have been doing a lot of damage since they set up a few days ago. We were ready to do it this morning until we saw you guys come on the scene. Rafael is trying to get to your guys before they all get wiped out. Charlie has probably been watching since you left Khe San; they probably have the ambush set. They might be watching you, actually the four of us, now."

"Did you say Rafael?" Parker asked.

"Yes, I did. He is one of our best. The only reason I use his name is because you know him. And he knows you."

Fres punched Mike on the shoulder and pointed toward a truck coming up the highway. It appeared to be Russian, which meant NVA.

"We've got to get down to your team. C'mon. Try and keep up. Fres will help you."

Mike disappeared into the brush. As the truck got closer to us, gunfire broke out from what I knew to be Collins' position. Parker and I followed Fres down the hillside, trying our best not to trip on the underbrush. I caught a glimpse of the NVA position on the highway, and noted that they were looking toward Collins, probably trying to find where the gunfire was coming from. One of them pointed a rifle at his position, then dropped as though shot.

An NVA machine gun opened up and began tearing up the hillside around Collins. Another rifle opened up near Collins' position and took out the NVA machine gunner. He was quickly replaced, and the new gunner continued to fire.

"That's Rafe," Mike said. "He's working alone. I have to get down there to help him out. Fres, see if you can get Parker and Aleck to Collins' position. If Collins has a radio, try to get through to our guys, and get them out here ASAP, ready to do some evacuations and some hunting. The weather is shitty, but it might be breaking up. See if you can raise a chopper from Khe San to help us out. We may have casualties."

Parker spoke up. "Wait. I owe Rafe. I promised myself if he ever got in trouble, I'd give him all the help I could. I'm coming with you. If you need me to lay in some covering fire while you guys do your magic, I'll do it. Morris is good in the bush, and he knows the weaponry. He can help Fres take care of whatever Collins is dealing with."

I took note of the truck and was glad to see that it was backing up, trying to get away from the firefight. That was good news; it said they were more transport than combat. It took Fres and I about fifteen minutes of bushwhacking to get down to Collins. Collins was out with a head wound. I looked for Pollock.

"We've taken some hits," he said. "Mary Jane and Hap are gone. Collins is functional but groggy. The only good news I have for you is the radio. I think Easterman has almost got it going again. We need some help, and the only way we can get it is the radio. If either you or your pal can get that radio to at least get some kind of a signal back to base, I'll put all you greenies in for some kind of a medal, or at least a beer back in your hooch in Lang Vei. Even you, Morris."

I had gotten to know Easterman on the way here. Whatever had happened to the radio was probably caused by rainwater getting into the works rather than any operator error. We found him busy drying parts and connections. Fres helped with this. In his time working with the Green Berets, Mike told me that Fres had made himself useful by working with the electronics. His English was still pretty rocky, but he was a fast learner, and he jumped right in with Easterman to dry parts.

Pretty soon they were able to get the radio to light up, but there was no real way to test the sending signal except to start sending a signal for help on the channels for Khe San and Lang Vei and keep repeating it.

Pollock came over to us. “I’m seeing some action from the NVA across the road. I think they are calling heavier firepower into our position. That’s bad news for us. Unless the shooters are able to take them out, our only hope is that your signal is getting out, and the white hats can get to us most ricky-tick. Is the radio working?”

“Looks good so far,” Easterman said. “We’re getting a signal out.”

A minute later, an artillery shell came over and blew up some trees about a hundred yards above us. Dirt, branches, and rocks started falling. Easterman spread his arms over the radio to protect it. The NVA were bracketing our position with artillery shells. A couple more shots and they would have us. Washington and Choi, the BAR man and the sharp-shooter, were scrambling to get to firing positions to the left and right away from the squad and the radio. Washington began firing the BAR in short bursts, while Choi concentrated on repeated aiming and firing with his sniper rifle.

I looked down and saw that Rafael and Parker were scrambling to cross the road, covering each other while one or the other tried to get close enough to use grenades and hand-to-hand knife work on whoever remained at Charlie’s position.

Somebody in the NVA position was calling in corrections to their artillery. The sharpshooters knew they would have to find him and take him out before he was able to guide the artillery to Collins and the radio.

***

Yesterday, I woke up groggy with my head and shoulder bandaged up. Not knowing where I was, I started trying to get out of the bed. A nurse came running up to me with some kind of a needle.

At least, I think it was yesterday. It might have been last week. Anyway, the nurse jabbed me with the needle, and I fell back to sleep. I feel a little more alert now, but I have no idea where I am or what happened to me. Last I remember, an artillery shell was heading my way during a firefight somewhere out on Route 9.

I am not in a regular hospital, though. I am in some kind of tent with several beds. Some of the other beds in my area are occupied by others with war wounds. I guess that this is a field hospital. I hear gunfire in the far distance, so I suppose we are near some action, but I don't know where. I can feel pain in several places on my body. The pain in my head is intense. My left eye is covered with a bandage. I can see and hear, though, and my thinking is clear enough that I am less worried about the severity of the injuries.

Most important, there are no bandages and no pain down near my private parts. Thank god. I have dodged "the wound to end all wounds." Scuttlebutt has it that a GI will give up any other body part to a combat war wound, but not those little babies between his legs.

Today, a female lieutenant with a stethoscope approached my bed.

"Hey, soldier. Welcome. If you are really awake, I'll bet you have a lot of questions. I'm going to take your temperature and check out some vital signs while we talk."

"Okay. First, I'm not a soldier. I'm a correspondent with The Prairie Observer, a little review of current events and news out of Minneapolis. My first question is who are you?"

"I'm Rafaela. I'm new on the job here, but I do have a medical degree and I do plan to do my residency for a real MD and a real hospital ER job when I get out. How in the world did you get yourself attached to a combat mission as a correspondent?"

"I had to work for it but going out with a squad on a Lurp is the most important thing I will do as a correspondent here. I promised the squad that I would write about it for the folks back home in a way that would help them understand what this war is all about. I can't believe my good luck in still having all my parts and most of my wits after getting wounded like this."

"Yes. You are very lucky."

Rafaela has beautiful eyes, dark skin, and beautiful dark, curly hair. More important than her looks, she has a way of looking at me that is totally engaging. With that bedside manner she is going to be a hell of a doctor. I discovered that I was getting a hard-on. I was torn between hoping she was not embarrassed by it, on the one hand, and, on the other, wishing I was mobile enough and knew enough about her to make some friendly moves. "Do you know anything about the other guys on my squad?"

"Yes, I do," she said. "How much detail do you want?"

"I recall that two were already dead by the time I got to their position. The squad leader, a sergeant named Collins, had been hit, and was pretty much out of it. Pollock and the rest of the squad were basically functional. Two sharpshooters were moving into better firing positions, but I don't recall them shooting at anybody. I was working with the radio operator, Easterman, trying to get a signal back to base for some help. There was a Hmong by the name of Fres Thao also working on the radio.

"Fres was working with a black Green Beret named Fredrickson who helped me get out of a bad situation when I first came up here from Saigon. There was a lieutenant named Parker who had been helping me out, and was with me on the Lurp, but he went to help another Green Beret named Rafael who he owed a favor."

Rafaela's eyes grew wide at the mention of the name. She became short of breath, and her talk was suddenly rushed.

"Is something the matter? I asked.

"Rafael is my cousin. I'm actually here because I am trying to find him. He has been out of touch for several months. All we can get out of the military is that he is on an extended confidential mission. You might be the first person I've met here who might have actually laid eyes on him."

"I'm sorry I don't know more." Actually, I did know more, but I was reluctant to tell this particular nurse anything about it. "I'm sorry, I don't remember anything after that except the noise of an artillery shell that was headed my way."

Rafaela had finally managed to relax as I related this information to her. "That's okay," she said. "You have given me some hope that he might still be alive. I have to caution you, though, about saying anything about Green Berets or their operations in this area. I don't think their presence here is a secret, but the things they do on some of their missions are. Just be discreet. I'd like to talk to you more, but I've got to look in on some other patients. I'll be back."

She turned back to me: "By the way, you are in a temporary field hospital at Khe San. The weather and enemy artillery and other actions are preventing evacuations for now. We hope to get you and others out of here to Da Nang in a day or two. Your friend Easterman is two beds down. He hasn't returned to consciousness yet, but he will soon. If you both like cribbage I'll see if I can rustle up a board."

"Just one more question. Do you know how I got here from the firefight?" I asked.

She looked impatiently at her watch. "No, I really don't. All I know for sure is that five days ago one of the air mobile helicopters came in hot with some damage from ground fire. I was told by one of the other nurses that you and Easterman were on the floor of it, unconscious. Between the weather and hostile fire going out and coming back, according to the pilot and gunner, you were all lucky to get here at all."

"How about the others?" I asked.

"I don't know. Sorry. The pilot went out again, but he didn't come back this time."

"What happened to Fres? Do you know?"

"There was no mention of any natives, Montagnard, Hmong, or other-wise, being involved in that particular action."

I thought for a moment as Rafaela turned to go again. "Wait," I said. She paused. "I don't know how the battle up on the highway turned out, so I don't know if I should tell you this." Rafaela turned back to look at me, her eyes wide and her look intense. "One of the last things I remember is Rafael and one of our riflemen moving in on the artillery setup that finally did my squad in. Those were heroic acts that I intend to report on. I wish I could tell you if they were successful."

Rafaela's voice quavered. "You were right to tell me. Thank you." She turned away from me then and left the room.

I have resolved to try and get a story about all this to Penny Blue as soon as possible. I need a typewriter, though, and I hope that Rafaela can find one I could borrow. I'm sure her operating room calm will return soon, but the constant uncertainty about her cousin Rafael over all these months has to be tough.

She returned later to change my "stop-the-bleeding" dressings into something less scary and more travel-worthy.

The next day a doctor came to see me. He told me my wounds were not particularly serious, but they would need attention over the next few days. "You are not confined to your bed, but you need to rest, and you need to stay close by. I don't want to have to go looking for you if we catch a break on transportation. Apparently the NVA is putting a major offensive together so I would like to get everybody out of here as soon as possible."

"Thanks for the update, doctor, but I left some good men up on the DMZ. Is there any way I can get back up there to see if I can help out?"

He shook his head. "You have to be crazy as shit to even think about doing that, and I have to be crazy as shit to even think about giving you a go-ahead. Don't ask me that again. Besides, you're a goddamn correspondent. You aren't even grunt infantry. You wouldn't know what to do with an AR-15 if it jumped up and bit you on the ass."

This didn't sit with me quite right, so I decided to set him straight. "Ah, yes," I said, sitting up as best I could on my hospital cot. "You don't under-stand, but I very much would know what to do with an AR-15 and with any number of other standard-issue small arms. So why don't you quit giving me so much shit about my going back up the road? You all have a firebase at Lang Vei. If you have a supply chopper or truck going up there, and the pilot has no objections, what are you going to do if I ride along?"

He wasn't backing down without a fight. "If you value your career as a journalist," he said, "you probably don't want to piss off somebody like me. I have a lot of important shit to worry about, and I am not going spend any time arguing with you about this. I am telling you, do not leave this compound without my say-so. If you do leave and you get killed, I'll try and get the commanding officer of whatever unit you are assigned to send a note to your mom. If you don't get killed, I'll give the MPs a full report of this conversation. You will be on your own to explain yourself."

With that, the doctor left. I knew in that very moment that it had become my sworn duty to get to Lang Vei to check out what happened on Route 9 a week ago, and, in particular, what had happened to Rafael, Army Green Beret and cousin to the beautiful Rafaela.

***

To get to the Lang Vei camp from the Khe San firebase I would have to go through the area where I was wounded. More importantly, it was the area where I had last seen Rafael. I had to decide whether or not to tell Rafaela about my plan, knowing that she would insist on going with me. When she asks, I will need to either have a convincing way to say absolutely not, or a hell of a good plan to convince her that we can get into the area and back out again without getting shot or captured. If Rafaela asks me what will happen to her military career if she is away from the field hospital when wounded soldiers are brought in I would have no answer for her, but I will have to discuss it with her tonight.

If I survive and can get back to Khe San more or less physically intact I have already written off any thoughts of another assignment as an in-country war correspondent.

The doctor did not restrict me to my hospital bed, so I decided to dress and head for the general mess where the off-duty grunts eat. Maybe I could pick up some intel on the current conditions on Route 9.

Once in the dining hall I took my C-rats to a table where five grunts were talking and playing a noisy card game with lots of cursing and slapping the table. Their rifles were stacked against an adjoining table, but the thing that convinced me that they were combat infantry were the Airborne shoulder patches on their jackets.

"Hey guys. Mind if I join you?"

"Shit yeah, man," one said. "Sit down. You got money or drugs?"

"Ah, no, but I do have a story to tell about how I got these wounds in a firefight on Route 9. Is that good enough?"

The hand slamming and bullshit ceased abruptly. All eyes turned to-ward me.

"Who you shittin', man? You're not even military. You're a correspondent hack trying to pick up some local color for some small town rag where everybody hates the baby killers, right?"

I shrugged. "Yeah. Basically, you're right about why I'm in this shithole country, but I'm not here in front of you hoping for some bullshit story for the home folks. I am looking for updated information on the conditions between here and Lang Vei. I got wounded in a firefight on Route 9 a week ago. I left a buddy back there, and I mean to go back and get him. Can you guys give me some help or not?"

"C'mon, man. Take the chip off your shoulder," the main instigator said. "Call me Hack. If you are worried about military rules you can call me Lieutenant Hack. You want a toke?"

"Sure. I'll take a drag." Hack passed what was left of a joint over to me. I took a long pull on the roach. I did manage to burn my fingers on it, but the charge felt good going down.

"So, you're saying that you were part of that Lurp that got tangled up with some NVA artillery a few days ago. Is that right?"

"That's right," I replied. One of the other grunts asked how I got assigned to that Lurp. "I earned my way by doing the push-ups and learning the mission and the armaments. Plus, I told the CO, a guy named Bukowski, that I would get his picture in his hometown paper."

Hack spoke up. "That's funny, man. I know Butthead. He's good people. So... if I ask Butthead how you earned your way onto that Lurp, he will know what I am talking about?"

"Yes."

Hack leaned forward. "Okay. What do you want from us?"

"Only information about the danger of traveling Route 9 between here and Lang Vei tomorrow morning."

"That's easy. The NVA are building up a big offensive. They've pulled their remote units back away from the DMZ to make sure everybody is coordinated when the attack starts. I think Route 9, today, is as safe as it has been since Dien Bien Phu in 1954."

I had to ask. "How do you know?"

"We were up there yesterday. It was quiet. What's your plan for getting up there anyway?"

"I haven't thought quite that far ahead yet, but I am all ears for anything you have to say about transportation opportunities."

"Well." He consulted with the other grunts. "Maybe there is a way we can help. Before you go much further with this dumbass idea, though, I have to tell you that I think the road will probably be safe enough for travel tomorrow. That means you stand a good chance of coming back alive." He paused for a moment. "But I wouldn't guarantee your safety beyond the day after tomorrow."

"What do you mean?"

"I mean that I don't know when the NVA will launch their attack. When they do launch, I think they will sweep south of the DMZ and we won't be able to stop them."

I couldn't believe I was hearing this kind of talk from a combat lieutenant. "Where the hell are you hearing this stuff? I'm not hearing anything in the news channels about an imminent attack—especially not one big enough to sweep everything south of the DMZ."

"If you believe something different than what I am telling you now then you are reading too many press releases from Westmoreland's PR office. You have to understand that Westmoreland may be the world's dumbest general. He gets his position by being the world's most handsome boozer and bullshit artist at Washington cocktail parties. He makes a great impression on the wives of politicians. Me and the other grunts keep trying to tell our commanders what we are learning in the field about NVA strategy. They keep telling us that our concerns are falling on deaf ears in HDQ. HDQ truly believes that using B-52s to bomb the Ho Chi Minh Trail will wear the enemy down, and he will soon come beggin' for a ceasefire."

Another GI spoke up. "They don't understand how fast Uncle Ho can shut down their operations on the trail, night or day. They can move their vehicles and people far enough away that they don't suffer much bomb damage. As soon as the bombers have moved on, they can quickly get the vehicles back on the road, fix any damage, and be on their way with their loads of machine guns, mortars, and ammo."

Hack spoke up again. "We believe that by now they have infiltrated several divisions into the south below the DMZ. At least a couple of those divisions are probably very close to Saigon. They are simply waiting for a signal from Uncle Ho that he is moving south across the DMZ. They will follow Ho till the end of time. Ho is the only leader they trust to throw out the invaders of their country: the Chinese, the French, the Catholics, and now the Americans.

"But you don't have to pay any attention to my ramblings. You've got a man down west of here. My guys here and me, we don't like the idea of leaving a man on the field after a fight under any circumstances, whether he is dead or alive. If your story checks out with Butthead, we will find him and bring him back."

I was speechless at this. "Jesus. My confidence that I can survive this has just gone up. I won't ask whether or not you are shitting me. Just tell me what you want me to do to make this happen."

"You need to tell us everything you know about that firefight before you got wounded. And you need to tell us what your real reason is for going out to find this guy. Was he a member of your Lurp, or was he a member of a different unit?" I must have hesitated because he said, "Come on, man. You got to tell us these things now."

"There are a couple of things I haven't told you," I admitted, "but here they are. This 'guy' is a Green Beret based at Lang Vei. His name is Rafael Barardi. I've never met him, but I know of him through a Hmong tribes-man named Fres Thao. Fres had been adopted by Rafael's unit as a guide and interpreter. They had trained him as a radio tech, but he had many skills and a lot of specific knowledge of the terrain, the people, and the territory. I was told that he had a lot of detailed knowledge of the Ho Chi Minh Trail over at least one hundred miles on either side of the intersection with Route 9.

"I know Fres because Fres got me through enemy lines when our helicopter took some ground-fire hits and had to crash-land several clicks below Khe San. This was a couple of months ago. Fres had crossed paths with our Lurp and was with us in the Route 9 firefight. He was working with another Green Beret sergeant named Mike Fredrickson.

"When I got wounded Rafael was trying to move up to take out the gunners who were killing us with artillery fire. I don't actually know if he got hit. He was still moving toward the artillery threat when my world went blank."

"Okay. I want you and I to go see Butthead. If Butthead tries to kill this recovery operation, then you and I will have to decide between ourselves how to proceed," Hack said. He waved toward his team. "These guys will have to make their own decision on this, but let's you and I try to figure out what we need to do next."

"There is one more thing I need to tell you," I said. "I have been debating whether or not I should, but you all seem to be willing to put a lot on the line for this and you need this last piece of the puzzle. Rafael has a cousin named Rafaela. She is a medical doctor who trained in trauma medicine at Case Western Reserve University in Cleveland. More important, she is here in Vietnam as a nurse. Actually, she is here on this base. She's the one who put me back together and made me able to talk to you now.

"I have gotten to know her. I like her. I want to help her find her cousin Rafael. I know she will want to come with us."

***

"You guys have got your heads so far up your asses you can see the cavities in your wisdom teeth. Hack. I told you to quit smoking that local shit. It will kill you."

Colonel Bukowski is in rare form today. Hack and I had laid the whole plan out for discussion. We needed a light truck, something bigger than a Jeep, and we needed to be ready for any NVA surprises along the way. Bukowski agreed with Hack's assessment of the enemy threat, but the situation around the DMZ was changing daily. The question was whether we needed something more than the standard weapons typical for a light duty patrol.

These questions were resolved in favor of somewhat heavier armaments packed away so they would not be visible, and therefore not seem threatening, from the surrounding hills. Hack and the three grunts I had met earlier in the day would ride in the back of our vehicle.

One question remained. We needed to talk to Rafaela's boss, the same doctor who told me that I could not leave Khe San. We needed to secure his permission for me to go, and because we would be looking for a man possibly wounded and possibly possessing important military information, we needed an experienced trauma nurse to go with us. Rafaela fit that bill perfectly. We just needed to get the boss doctor to agree. As a civilian it was clear that I should stay out of that discussion.

I counted myself lucky that Bukowski had gotten his head into the mission mostly because he had not heard anything from Lang Vei since the firefight on Route 9. In addition to the military mantra about leaving no wounded behind Bukowski also felt duty bound by his mission, his rank, and the battlefield history of the United States Marines to find out what was going on out there, especially given the rumors of an impending attack from the north across the DMZ.

Bukowski would talk to the doctor and get his permission to take Rafaela along.

Bukowski and Hack decided that we would leave just before sunup tomorrow morning. We hope to take advantage of the irregular light conditions so we can try to have a trouble-free ride to Lang Vei.

While Bukowski and Hack and the grunts went to get their equipment ready, I went to brief Rafaela and get her head into the mission. If she had concerns, I was to bring her immediately to Bukowski and Hack to discuss them.

***

"I was wondering where you had gotten off to," Rafaela said. "My super-visor told me you were free to move around, but you could not leave these facilities. Your wounds still need attention."

"I know," I said, "but things are happening fast, and you need to get ready to go on an emergency mission."

She was astonished. "What did you say?"

"I'll tell you as you start packing your gear. The unit commander here, Colonel Bukowski, has authorized a mission to find out what has happened at Lang Vei. He hasn't heard anything since our firefight, and he needs to figure out what happened. If there are casualties, we need to bring them back. If there are wounded, that's where you come in. How soon can you be ready?"

"I have to hear from my supervisor, Dr. Castro, before I am willing to stop doing my patient rounds, but I'll be ready if he assigns me to your mission."

"A big reason why you have been chosen for this is your training as a trauma doctor. Bring whatever equipment and supplies you need for that sort of work in the field. Can you handle that?"

"I can handle that. Now get out of here so I can finish my rounds and gather my gear."

I lingered in the doorway. "What if one of the wounded is your cousin?"

I could tell I'd struck a nerve. "Don't patronize me," Rafaela said. "Just get back to your business. When your team is ready, I will be ready."

I had finished packing my gear and laid down for a nap when Hack showed up.

Get up. We're pulling out in fifteen minutes."

"What? What the hell time is it?"

"It is early, so get up. Get up now, goddamn it!"

He gave me another rough shaking. As I started to roll out of the sack he said, "We are about to get a rain dump. Bukowski thinks we can take advantage of the cover the rain offers and cut down the risk of engagement if we leave now. We might be able to make it all the way to Lang Vei before the rain stops."

"Is everybody signed off on this? What about Rafaela and her supervisor?"

"Everybody is signed off. Bukowski wants us to go now, and he means right now. Do you have your gear ready?"

"Yeah. I'm ready. Except for my rain gear and long johns. Give me a minute."

"We will be out front. We've got an armored six-by-six with a canopy. It will be like riding in a limo," Hack said.

I came out a few minutes later. Hack helped me throw my junk into the back of the truck. Then we jumped in and took off, heading for Route 9.

Rafaela was in the back with me and three of Hack's grunts from the mess hall. They were checking our armaments. We had a .30-caliber and a .50-caliber Browning machine gun plus a variety of personal weapons like .45-cal service pistols and AR-15s with boxes of ammo. We even had a couple of M-1 rifles with bayonets.

So much heavy shit was happening in such a short time that my head was spinning. Hack had crawled up front with the driver, who was another member of his team.

He turned to us. "You guys had better make sure those weapons are functional for you. If you don't know how to load, aim, and shoot at least one of those weapons, you ain't gonna be worth much to anybody here including yourself. My guys will handle the heavy stuff. Aleck, Rafaela? Can you guys handle those light weapons?

Rafaela answered. "Yeah, Hack. I was raised on a farm. I can take a rabbit's eye out at a hundred yards. What else do you want to know?"

"No shit. I believe it. How 'bout you, Aleck?"

"I got training from the Lurp guys and plenty of weapons experience the last time I was on this road. That was about a week ago. I don't think more training will help me at this point."

"I'll take your word for it. We're deep in enemy territory now. If it looks like we are going to have enemy contact, the main thing is to listen care-fully to what I have to tell you. Basically, it's this: keep alert, be prepared to move fast, and stay behind me or whichever of my guys you can see. Do your best to keep up. If you lose sight of us, get down on the ground and stay there. One of us will come and get you as soon as we can."

"If one of your guys comes to find us, how will we know he's a good guy?" Rafaela asked.

"He will be singing a quiet version of 'Take Me Out to the Ball Game.' It'll sound more like a bird than a human, so listen carefully if you hear any bird sounds." He demonstrated. "You may have to announce yourself to him, so be careful. You won't want to announce yourself to the enemy, too. That's enough questions for now. We should be on the Lang Vei com-pound in about fifteen minutes. I count it as good news that we haven't had any trouble so far.

"I have to go back up front to help my driver find the compound. When the truck stops, look at me. If I hold my hand up with all my fingers out, then make a fist, get out the back of the truck fast and come around to the front so I can tell you which directions to move in. Got it?"

We both nodded.

Hack was true to his word; I estimated the time between our conference and the truck stopping at very close to fifteen minutes. The rain had increased, and the driver had buried us deep in some brush off the road. When Hack gave us the signal, we jumped out the back and ran around to the front with the three grunts who had been manning the heavy guns. The driver jumped into the back. He would let us know by radio signal if Charlie came into the area. If worse came to worst, he would unlimber the machine guns and try to keep the path back to the truck clear for our return.

"The main building in this compound should be about fifty paces ahead," Hack told his men. "You three start moving in the direction I indicate with my flashlight. When you get to the building don't go in. Take a quick look around, and one of you come back and get me. If it looks clear, we will move up."

"Yessir."

Within a few minutes one of the men had returned. "Hey, cap. Looks clear to us. It is pretty quiet so far. We should all move up now."

"Okay. Rafaela. Aleck. Let's move."

The building was quiet as we approached. We stood back about twenty feet. The point man touched the door. It was unlocked. He pushed it open and stepped in, where he immediately appeared to trip and fall. The door slammed shut behind him.

"Down! Now!" Hack screamed. "Whoever is in there, we are Marines and US Army Special Forces and we are well-armed. My men and I will destroy this building and everything in it unless you identify yourselves right now."

A voice with a heavy foreign, possibly Vietnamese, accent came from within the building: "You are army?"

"Yes. Can my man speak?"

"Yes. He speaks now."

"It looks okay to me, Cap," the point man said from the other side of the door. "More a misunderstanding than anything. It looks like there are two wounded men here. This guy sounds Hmong to me. He may be caring for these guys. They look pretty rough. You need to send Rafaela in here right now."

Rafaela spoke up. "I'm coming in now. Open the door for me."

"Wait," Hack whispered. "I want to open that door. Aleck, you come with me. Stay close behind. If that guy is Hmong, you need to try to ID him. One of you guys needs to cover us and keep the doctor behind you. The other needs to go around behind the building and check it out."

In my opinion, time was of the essence. I decided not to wait. "Fres! Is that you?"

From inside the building, I heard, "Aleck Morse? Is that you? How'd you get here? That chopper that took you and Easterman out of here didn't come back. I figured you all crashed and were now dead."

"Nah. It takes a lot more than that to knock out a guy from Duluth. Long story. Hey. These army guys with me are still worried that we might be getting lured into a trap. Can you come out the door with your hands up? I'll be just outside to greet you. There will be a lot of weaponry pointed your way when the door opens. Just be calm."

"I trust you, Aleck. Did someone say Rafaela? If she's with you tell her cousin Rafael is here. He is wounded and needs help right away."

At that point Rafaela's trauma-doctor cool collapsed. I heard a low moan behind me followed by deep, racking sobs while she tried to catch her breath and move toward the now open door that framed Fres with his hands up. She did not wait for any niceties. Hack moved out of her way.

Where is my cousin!" she demanded of Fres.

"I will take you to him," he said.

"Aleck bring the rest of my gear! Right now! Hack! I may need a stretcher for both of these wounded men. I don't know what I am getting into here, but I want to be ready to move them as quickly as possible back to the truck. I need sturdy stretchers that minimize the possibility of making their wounds worse. Can you handle that?"

Hack nodded. "My men and I will cut a couple of small trees that we can lash together. Fres—is that your name? Help the doctor get set with whatever she needs, then help me find something to lash the trees together, some rope or twine or string. Even vines from some of these trees will help."

Fres showed us into a back room where the two wounded men lay. It was not immediately clear whether they were unconscious or dead. Rafaela went to her cousin and began hugging and kissing him while her tears dripped onto his dirty face and found their way through his ragged beard. At the same time, she was feeling for a pulse and other vital signs. She carefully removed the dirty bandages covering his wounds.

"Aleck," she said, furiously wiping away tears and focusing up, "it may be impossible but search this building for any medicines or disinfectants, clean towels, and any source of water for cleaning these wounds. Have Fres help you look while he tries to find rope or string for the stretchers."

"Anything else?"

"Yes. First, if you know anything about first aid, please check the other man out and see if you can figure out what we need to do to get him ready to move to the truck."

Fres spoke up. "He's one of your recon team, Aleck. He took a bad blow to head. Knocked his helmet off and gave him a bad cut. He has been out since the firefight, but still breathing. He seems normal in any other way except being awake. Pulse and heart rate normal. Not much bleeding except for the head wound."

The other man was Collin's sharpshooter, Choi. "How in the hell did you get them here from the firefight?" I asked.

"I hid them until the battle ended and the last NVA soldiers left. Then I carried them here," he replied.

Rafaela seemed overwhelmed by this man's courage. "Thank you, Fres. I want to hug you and kiss you, but that will have to wait. You deserve a medal. Go ahead now with Aleck to look for medical supplies and rope. I'll check the other man as soon as I can get my cousin stabilized and ready to move."

Hack stuck his head in the door. "We need to get out of here fast. Make sure your wounded men are ready to move in five minutes. Got that?"

"Yes, I have it. I am working as fast as I can," Rafaela replied.

Hack went back to help with building the two stretchers. When Rafaela said they were ready to move, Hack and his men took the stretchers into the building and then moved each of the men back to the truck. They lay the wounded onto some padding that had been brought in the truck for that purpose.

"This is going to be a rough ride," Hack said. "Stay low. This truck has some armor plating along the sides that should help protect you against small arms fire as long as you keep your head down and we keep moving. Don't raise your head unless we get stopped or I tell you to. If there are any NVA along the road with heavy weapons we may be out of luck. If Tommy needs any help with that Browning .50-cal, Aleck, be prepared to help him.

"If we are attacked by anybody with serious intent to take us out, we are all going to be busy. The main idea here is to protect the doctor and the two wounded men."

The driver backed up over the same short path that we plowed down when we first turned off the road. Tommy, the machine gunner, and I had a hard time keeping our heads down. We wanted to be the first ones to see if any problems in the form of hostile enemy showed themselves. Once the truck was back on the road, the driver threw it into a forward gear and jammed his foot on the gas. So far it looked like we hadn't attracted the attention of the NVA.

I began to worry as the truck kept picking up speed over the bombed-out road. We were hitting some bad bumps, and the wounded men were getting bounced around. I certainly didn't want to have an accident because we were going too fast over bad roads. Rafaela was using her own body to cushion her cousin against the bumps, and I knew she was going to get badly beat up in the process. Her eyes caught mine and looked toward the driver. She was pleading to have the driver slow down.

"Hack!" I shouted. "Tell your driver to slow down! The bumps are killing these guys back here."

I remembered the driver telling me that this particular version of the truck would not go over forty miles per hour so I relaxed a little, but I still could not bring myself to keep my head down. I had to know what was going on around me; perhaps something or someone was hidden in the brush along the side of the road. Tommy's head was also on a swivel.

Hack motioned to the driver to slow down. He did, briefly, but then I heard some shots ring out simultaneously with a couple of pinging sounds on the sides of the truck. Somebody was shooting at us.

"Get down back there and stay down!" Hack yelled. "We are taking fire."

I had been trying to help Rafaela comfort her cousin while the other two GIs tried to protect Choi from too much banging around. Now that we were getting hit, both of them had to work to get the Browning .50 ready to return fire.

Rafaela was on her side with her arms and legs around her cousin. I had been on the opposite side facing her, our arms and legs occasionally entangled. In the pauses between bumps our eyes locked as we tried to prepare for the next jolt. When the next bump hit she shouted in pain then quickly bit her lip and looked away from me. Tears were streaming from her eyes. I heard a low moan build in her.

By now the two GIs had abandoned Choi as they worked to get their machine gun ready to return fire. Choi was bouncing around like a spastic marionette. Whatever his original injuries, Choi's bouncing was producing life-threatening injuries of its own.

I looked at Rafaela and nodded my head toward him. She nodded back at me. I went to Choi and, like Rafaela, wrapped my arms and legs around him. The bruises I had gained with my arms around the cousin were even more intense now as they were joined by new ones. With each bump I wanted to scream as loud as I was capable of, and I'm sure I did more than once.

The GIs had the Browning unlimbered, loaded, and aimed out the back of the truck, but no more shots were fired at us. After about ten more minutes on the road Hack told the driver to slow down. We were getting to the outer security perimeter at Khe San. Hack told Rafaela to check both wounded for wounds that might need some kind of emergency response beyond the standard protocol so that he could call in the trauma staff and tell them what to prepare for.

He had already alerted base ops to prepare for our arrival and not to shoot. “Clear our way to the trauma tent and give us an escort. We will approach at top speed with our headlights flashing SOS,” he said into the radio.

Rafaela was hurt from all the pounding. I prayed none of her injuries were serious. I was feeling pretty rugged myself, but I felt I could still function. “Do you want me to take some notes while you examine these guys?” I asked.

Her mouth formed a rictus of grimace, quickly gone, as she acknowledged my question. Her voice was stressed as she answered. “Maybe you can help me with the instruments and the undressing.”

After a few minutes she and I could report that no extraordinary procedures were required, but she did want to emphasize, again, that Rafael’s wounds to his head, left shoulder, and leg did not appear to be life-threatening. However, his coma condition, and that of Choi since the earlier firefight, were of great concern to her. She wanted a trauma specialist with the right experience in Khe San as soon as possible. If flight ops were possible any time there was a break in the weather, she asked that an emergency evac flight be initiated ASAP.

As soon as the trauma staff had Rafael and Choi out of the truck and into the trauma tent Rafaela appeared to pass out. I rushed to her side and bent toward her to see what was wrong. I whispered into her ear. “Rafaela, Rafaela. Talk to me.” I waited. “Talk to me.” There was no response.

I pressed my ear to her mouth. Her breathing appeared to be normal, if a little ragged. I pressed my hand under her shirt to feel for a heartbeat somewhere around her left breast. I felt her wrist for a pulse but was not sure of what I was feeling either around her breast or her wrist. I began to check her body for blood, or for any signs of bruising that could cause pain. Her head and face were bruised because of her attempts to cushion her cousin against the heavy bumping along Route 9. Her body might show signs of other trauma, but I did not feel competent to judge what I might find.

I carefully rolled her over but could not find anything obvious that would explain her condition. There was no blood on her clothes that told me she had a bleeding wound, but I was reluctant to undress her to check further. Two medics were waiting for her outside the truck, so I called to them to get a stretcher and help me get her into the trauma center.

I was beginning to feel a little faint myself. My own wounds from the earlier firefight with the Lurp team were not close to being healed, and I probably had pulled some of the stitches in my leg and head wounds during this rescue mission.

When the two medics came back with two other GIs, I did what I could to help, but I knew I was seriously close to passing out. I moved aside and told them I might need their help when they were through moving Rafa-ela into the trauma tent. One of the medics yelled for another stretcher. I could see a vision of Dr. Castro in my mind's eye. I could imagine how pissed he was going to be about all this.

That is the last thing I remember.

***

I woke to Dr. Castro shining a bright light into my eye. I jerked my head away. "Where the hell am I? What the hell is going on?"

Dr. Castro was obviously pissed. "Where do you think you are? You goddamn kids in this goddamn war. What the fuck do you think you, a goddamn wounded civilian under my care and a goddamn US Citizen on top of it all, were doing?!"

"Where is Rafaela?" I asked, calmly.

"She is in the operating theater getting checked out. She looks okay to me, but she is beat-up and bruised. You're next to go in there. What the fuck were you guys doing out there to get so banged up? Did you have a goddamn boxing match?"

"Didn't Bukowski say something to you about what we were doing?" I asked.

"Sure. He said you and Rafaela needed to identify and help some wounded guys that you both knew."

"Did he tell you where we were going?"

"No. Where did you go?"

"Gosh, doc, I don't know what is classified info and what is not. You'll have to ask Bukowski."

"Don't call me doc, you little twerp."

With that closing remark, Dr. Castro left the room.

Soon they rolled Rafaela out of the operating room. She asked them to stop by my cart. She reached out to my arm, then leaned toward me trying to kiss me on the cheek. Obviously, she was still in pain. "

Thank you for everything you did," she whispered. "You saved my cousin's life, and I want to repay you somehow. Talk to me later," she said as they rolled me into the operating theater.

Her hand and fingers slid across my arm, as though she didn't want to let go just yet. I wanted to look after her, but a sharp pain in my neck brought me back into the present moment.

"I am in love with this woman," I said to myself. "If I have my way, I will talk to her every day for the rest of my life."

END

Tree of Life

Early fall, sometime in the near future. Ticasuk Vitti is a crab-pot puller on a crab boat with an all-women crew. They are working the Bering Sea crab fishery near St. Paul Island, Alaska. In recent weeks marine radio has been reporting increasing hostility in the public exchanges between US President Trump and North Korea's Supreme Leader Kim Jong Un. Ticasuk, nicknamed Tica, fears the possibility of nuclear war in the south. She asks the captain to let her off at St. Paul. She will try to find a pilot who will fly her the seven hundred miles north to Kotzebue, then to her mom's home above the Arctic Circle near Ambler on the Kobuk River.

PART 1

When Marian Delaney and her crew first hear the news over their boat's satellite wi-fi, they chalk Trump's public insults to the leader of North Korea as more political posturing; a pissing contest between two ego bound children playing with some dangerous toys. There aren't many of those kinds of pissing contests on an all-women crab boat, so they roll their eyes and continue with their responsibilities.

Ruggles has the wheel of the 140-foot Aleutian Challenger, a commercial crab boat working the crab fishing grounds three hundred miles north of Dutch Harbor. "Dutch" is the prime port for the Bering Sea fishing fleet. It is located halfway down Alaska's Aleutian Chain near the city of Unalaska. The Challenger, under the command of forty-two-year-old Captain Marian Delaney of Seattle, has been working the Aleutian fishery for the past few weeks.

Delaney knows her load of crab from this trip will probably be meager. Nobody on the boat wants to think or talk about global warming as a cause of changes in the health of the crab stocks, but there is little doubt that the rapidly shrinking ice pack in the Arctic Ocean will cause the Arctic Ocean itself to warm rapidly. There can be little doubt that the Bering and Chukchi seas will also warm rapidly. The warming seas are creating new and powerful changes in the ecology of the crab fishery. The future effects on the commercial crab stocks or any of the commercial fish stocks in the north are unknown and perhaps unknowable.

The crew of the Challenger are not alone in ignoring the evidence of their senses when confronting the effects of global warming on their living and livelihood. With a general political and social trend, these days, that condemns science, scientific inquiry and knowledge itself the effects of the rapid changes on the eco systems of the north may never be known until it is too late to make corrections.

For this season Delaney and the boat's owners, Tom and Marjorie Kettleman out of Kent, Washington, have decided to keep the crew and the boat in good fishing form, just in case the fishery has improved this season at least in those places where they decide to set their pots.

Today, the skies are clear, but the winds are strong, and the seas have been roughened up with twenty-foot waves and white caps. In heavy seas Delaney prefers to have a skilled and focused crewmember at the wheel, even with an autopilot, and of all the crew Ann Ruggles is the most skilled at this. But constant focus on rough seas, the wheel, and maintaining a constant heading is tiring, and it is time to give Ann a break.

"So far," Delaney mutters to herself, "there is not much magic in this fishery." Louder, she asks, "Are you ready for a break, Ann?"

"Yeah, cap. My arms and legs are starting to twitch from fighting the wheel and the rolling."

Marnie Thompson, the boat's engineer and Delaney's second-in-command comes up the stairs from the crew lounge with its small library of books and DVDs. "Here's some coffee, Marian." Beyond her skill with engines, navigation systems, and slippery decks Marnie is also an attractive woman at ease with herself. She and Captain Delaney have been friends and professional colleagues since meeting at a Coast Guard workshop on safety procedures a few years ago. Delaney was immediately impressed by Marnie's self-confidence and her impressive academic record in math and mechanical engineering from Brown University.

Delaney had been first officer on the Challenger for two seasons before meeting Marnie. As soon as Delaney had all of the necessary licenses and the captain's chair became vacant, she proposed to Marnie that they pitch themselves to the boat's owners as a team to operate the Challenger during the season. The owners accepted eagerly.

In the Aleutian fishing community, the mutual respect between the two women has sometimes led to speculations about their sexual orientation. To those who ask direct questions of either their response is usually a polite variation of "Get a life." Among the boat crew the dockside rumors are ignored.

Delaney takes the cup and sips the hot, sweet liquid. Sadie, the cook, fixes it just the way she likes it: lots of sugar and cream. If she puts it on the table there is a chance the boat will roll enough to cause the cup to slide off and spill everything. "Thanks, Marnie." She holds onto the cup with both hands. She lets the warmth flow into her palms and fingers and wrists.

"How are the engines doing, Marnie?" she asks. "Can you take an hour or so at the wheel? Ann needs a break."

"I can take over for a little while, but I need to be listening for any chatter over the PA from Insh down in the engine and transmission spaces. I've got her keeping a close eye on a couple of gasket problems that have gotten more worrisome. If one of them breaks I will have to go down there right away to help install a new one."

***

At dinner that evening, while Ann Ruggles maintains the helm, Delaney announces her plan. "I want to do one more small set before we go to St. Paul to fuel up. I think there are some crab in a little depression on the edge of the shelf about seventy miles southwest of St. Paul. I'd like to do a set of twenty pots before we quit. Once we do that set, we'll head for St. Paul to unload and fuel up. Do any of you have any questions or concerns? Any bitches about the trip or my handling of the boat?"

Insh—officially the engine room's oiler; unofficially, the only person on the crew who knows the details of every mechanical and engineering system issue on the boat—speaks in her Irish brogue. "Nah. Yer doin' good, cap. Though I wish you'd got us more crab. My Guinness budget is all shot to hell after this shitty season."

"I doubt if the seasons ahead look much better," Marnie says, "but maybe they will after the crab stocks have had a chance to rebuild."

"They might get better, that is, if the ocean stops warming, or if the Russians and Chinese don't get them all first," Delaney says.

"Or if Trump and Kim Jong Un don't blow us all away while they are playing chicken with their little nukes, "Insh said.

Sadie, the gray-haired, stocky cook, chimes in. "I'm with you, Insh. I'm going to have to get a job in Dutch flippin' burgers this winter. I can't even afford to get to Anchorage to see my buds at Chilkoot Charlie's for football and beer."

Delaney is not sure about Sadie. She's only had two fishing trips on the Challenger, and she has not really been tested on the deck during heavy seas. She's a good cook and loves cooking for this crew, but there does seem to be an edge to her attitude if she is under any stress. Though she does not make an issue of it, Sadie wears a Saint Christopher medal, and she carries a pocket version of the Bible. If something comes up that requires Sadie's attention, Ann Ruggles knows where to look for her. There are only a couple of quiet places on the boat suitable for hiding out for the purposes of reading.

"I'd like to get off at St. Paul," Tica Vitti says. Tica is the newest and greenest member of the crew. She was born almost twenty-two years ago to her Inupiaq mother, Mary Lincoln, and Italian father, Antonio Vitti. Small, slender, and extensively marked with traditional Inupiaq face and body tattoos, Tica did not at first appear to Delaney to be a good prospect for hard physical work on the Challenger, but she had a good reputation around Dutch Harbor's shoreside community. She had some good skills with electronic systems, but she also took on hard jobs like cleaning the barnacles off the bottoms of boats. She worked hard to get her share of the work done, and by various means she could persuade others to do theirs.

Like her mother and many in her home community of Point Hope, Tica is a talented visual artist. She came to the Challenger with a sketchpad and a small pouch of colored inks, pencils, charcoals and erasers. She often makes quick sketches to use as trading stock with crewmembers who might have something of value, like easy or interesting job assignments, to exchange.

On a crab boat like the Challenger, there are constant tasks and odd jobs to be done as assigned daily by the captain. Tica has discovered, for now, that her sketches have value in purchasing the time of others to do the chores she has been assigned but does not want to do.

"The work of this fishing trip will be done after we unload at St. Paul. I'm worried about the war talk, and my mother is probably worried, too. She will need some help if things start to go sideways in the States. If I can use the satellite phone once we are done with the set, I can ask my friend with an airplane to come down from Kotzebue, pick me up in St. Paul, and take me up the Kobuk to my mom's place."

Delaney looks over toward Marnie Thompson. Tica's request is a surprise, and Delaney does not know how to respond. Tica has been a good worker on this trip. She works hard and showed enthusiasm, energy, and initiative when learning how to deal with dangerous and tricky procedures. She is tolerant of some of the rough talk and teasing among the crew about her facial tattoos. She tried to join in their verbal roughhousing whenever she had something to say.

"Do you see any problem with that, Marnie?" Delaney asks.

"No, Marian. I'm okay with it. I'm getting a little concerned about the war talk myself, but there is no way I can go back to my family in Douglasville, Georgia, even if I wanted to. Can I come up the Kobuk with you, Tica?"

"We can talk about it, if you're serious," Tica says.

"I might be serious. I'll think about it. Thank you."

Delaney speaks up. "Don't get too carried away with that idea. I realize we don't have that much of a load, but we are three hundred miles north of Dutch, and I'm not very enthusiastic about being short a first mate and a hard-working crew member on such a long run. A lot will depend on the forecasts for weather and winds, but this time of year, as you well know, Marnie, weather reports can be pretty unreliable. If I judge it to be necessary for whatever reason, I won't let either of you off the boat. Are we understood on this?"

The two women nod.

"Weather is moving in in a few days," Delaney says. "I'd like to get to the first set by dawn tomorrow with the first recovery twenty-four hours later. We should be able to get back to St. Paul by sundown in two days if we continue underway," She looked around the room. "Any problems? If not, Ann and I will take turns in the wheelhouse; Marnie and Insh in the engine room; Sadie and Tica in the rack to get some sleep so you can relieve us in a few hours. We can catch up on sleep after we set the pots tomorrow morning."

***

Tica cannot sleep. She lies quietly in bed lightly rubbing her face. After many years she knows by now how her tattoos lie on her face. She is very proud of them and of her aunties in Point Hope who laid them on. As her fingers trace the various lines in her tattoos, she takes comfort in their patterns and in their connections to the animism of the Inupiaq. This is her form of bedtime prayer. As she traces each one, she wishes for the loving touch of the women who applied each of the cuts and coloring ink and ash that produced both the pain and the beauty of the tattoos; all this while they hummed the ancient songs.

Tica's bed is near the Wi-Fi hot spot in the bulkhead, so when she still cannot sleep, she lies there scrolling through Facebook under the covers.

She taps on an article: "Ten Cities in the US that Could be Hit by North Korea ICBMs." She hates the writer for using such a blasé title, and she hates herself for being interested enough to click on it. Even so, she's worried about her mother. Tica feels hopeless out here in the strait, where she can't do anything to help. The targets listed include the usual suspects, Honolulu, Los Angeles, Seattle, etc. When she sees that a Ronald Reagan inspired Star Wars anti-missile base near the Alaska High-way community of Delta Junction, 100 miles southeast of Fairbanks, is in the top five, her blood runs cold.

Tica knows that even a miniscule error in the trajectory of such a missile could drop an incoming missile on her mother's community of Ambler. With the thought that a similar error could drop the missile on her home community of Point Hope, her eyes began to burn and tear up.

Desperate, she casts her mind onto an item of lesser stress, stress she can do something about, the crab pot recovery.

Her mind churns over memories of her role in the earlier sets and recoveries of crab pots on this trip. The pots are big; they are heavy and bulky. They are made out of steel and can weigh up to six hundred pounds empty. When they are loaded with bait and enough rope to reach as far as a thousand feet down, they can weigh up to seven hundred fifty pounds.

Immediately behind the wheelhouse and superstructure the pots are loaded, unloaded, and moved around by a hydraulic crane permanently fixed to the boat's hull. The crane is used to move an electromagnetic pad that connects to the pots so they can be lifted. For the last set a week ago Marnie assigned Sadie the job of preparing the bait on the deck, then baiting the pots before they were lifted over the side and dropped into the ocean depths.

In that set, Marnie assigned Tica as location recorder. She worked in the wheelhouse with the captain to record Pot ID numbers and GPS locations when Marnie gives the signal from the deck that the pot is resting on the ocean floor. Tica enjoyed the job of recorder and hopes to have the same assignment on future sets.

Cap has said, though, that her policy is to give everybody operational time in every position on the boat. In her view the crew rotation policy best assures the boat's safe and profitable return to the dock after a trip. It also provides an individual feeling of security in trusting her own skills and judgement in any position on the boat, and it provides a sense of fairness among the crew while the trip is underway that no one is immune from an assignment to clean the toilets.

Tica is convinced that the coming set will be her time on the deck. She can only pray that the seas would be calm. On a set, the intricate movements by the deck boss and pot wrangler, usually Marnie; the shot and bait wrangler, Sadie last time; and Insh, working closely with Marnie to make sure the ropes follow the pot cleanly, form a kind of ballet.

If things go well, and the seas are relatively calm, the time required to bait and set a single pot might be between two and three minutes. If the seas are rough, the potential dangers to the crew go up exponentially, as does the time required for a set. Loss of life, or serious injury among crab boat crew members has come down over the years because of tighter rules imposed by the Coast

Guard about safety equipment, procedures, and training. However, each set or recovery, rough seas or calm, is fraught with high risk to all members of the deck crew if one of them, for whatever reason, gets careless.

Though each member of the crew wears a headset and mic during set and recovery, Delaney's policy is to rely on hand signals. In that way the deck crew will always be able to communicate with each other and with the wheelhouse even if the Challenger's electrical system or shipboard Wi-Fi network go down. Delaney's policy has the added benefit of keeping the signals simple and clear and to the point. A big problem with the earphones and personal mics is the temptation to gossip and shoot the bull. Delaney wants that kept to an absolute minimum as long as pots are moving around the deck.

The recovery will be even more dangerous than the set because the pots will weigh much more with their load of crab. A successful "set" can bring with it a lot of good news for the crew in the form of abundant, mature, and legal live crab, but there is always a dark and lingering possibility, at any moment during recovery of a pot heavy with crab, of some sudden accident because of inattention or misstep by a crew member, or, worst of all, by a sudden mechanical failure combined with a rogue wave.

These possibilities weigh heavily on Tica's mind. She is restless. She wants to get this last haul over with so that she can get home. She doesn't fall asleep for a few hours. Sadie, on the lower bunk, snores contentedly, as though she is drugged.

***

"Don't linger too long over breakfast," Delaney says the next day. "I want to wrap up the recovery by lunchtime tomorrow so we can get back to St. Paul and tie up before the storm hits. Marnie, you are on the deck with Insh and Tica. Ann, you are in the wheelhouse with me. Sadie, you will stay in the engine room. I think you have learned enough to know the difference between a growing problem and a growing catastrophe. If you see something start to go seriously wrong get Marnie or Insh on the phone. Better yet, stay on the phone with them from the start. Check with them both every couple of minutes to make sure the phone connection is still okay.

"If the phone goes dead, and the situation looks serious to you, go up to the deck, fast, and get somebody's attention. If they are too busy handling pots, come up to the wheelhouse and tell me what's going on. Once Marnie takes control of the power block, she won't be going anyplace else unless I tell her to. Talk to Insh first. Tell her what is going on. Can you handle it Sadie?"

Though Tica is nervous about her own assignment, she can see that Sadie looks shocked. Her face is pale. Wide-eyed she asks: "Are you sure, cap? I don't think I'm up for this. I was going to cook up a good breakfast for you guys after you complete the sets. I know you like to shift people around, but Tica is not smart enough to know how to bait the pots and count the shots. And all I really know about the engine room is where to grab a fire extinguisher if a fire breaks out."

Delaney responds. "Insh and I will take a quick walking tour of the engine and transmission spaces with you as soon as we finish here. We will refresh your memory about the most important gauges and valves to watch. It will take us about thirty minutes to get to our first set, and we will answer any questions you have while we get underway."

The seas seem to be less rough, but the boat still rolls a lot during Delaney's briefing. Tica is red-faced and breathing heavily.

She does not want to look at Sadie. She feels like Sadie has put her down in some way. She tries to hide it, but her voice shakes as she tries to express her deep concerns about being on the deck with a real job to do during either a set or recovery. "I appreciate this opportunity, cap, but I don't have enough confidence in my abilities or experience to feel that I will be able to deal with anything out of the ordinary."

"Marnie and Insh are going to be right there with you," Delaney says. "They both recommended you to me for this assignment; they said you are ready for it, and you need the experience.

"I'm reasonably confident that the seas will be calm throughout the set, but we can revisit this again before we start recovery. If the seas are still too rough, I might consider new assignments depending on what Marnie has to say. Don't count on it, though, Tica; I know you can do this no matter how rough the seas are, and no matter what Sadie says."

Tica takes a deep breath and glances in Marnie's direction. Marnie winks at her. Insh gives Tica one of her specialty crooked smiles, full of teeth, mischief, and meaning.

Tica turns back to her breakfast plate and piles on some more eggs, bacon, toast, and potatoes. She takes a blueberry pancake on the side and slathers it with butter and maple syrup. Though she is slender, she eats a lot because that is the way her family brought her up. Like the bears in their den, the Inupiaq consider that loading up on fat and carbohydrates, usually oily seal meat or a chunk of whale skin and blubber called muktuk rather than pancakes, is a simple matter of surviving the winter in good health.

In the white man's world, and the Challenger is very much a part of that world, as much as Tica tries to fight the temptation, she usually turns to comfort food when stressed.

Delaney brings the coffee pot over and squeezes Tica's shoulder. "Re-fill?"

Tica smiles up at her. She refills the cup and then leaves the lounge to go up to the bridge to check with Ann on preparations.

Marnie walks over. "We will be underway in about thirty minutes. Insh and I want to spend a few minutes going over procedures on deck with you before we move up to the first set. Hurry up and finish your break-fast. First, we'll take a walk around the deck."

After clearing her plate, Tica puts on her bright yellow float coat, safety harness, and waterproof gear and boots then steps out onto the pot deck. The Challenger's superstructure blocks the worst of the winds. They are calm here, just outside the door to the storage spaces.

As Tica moves farther out on deck, though, the wind begins to whip around the stacked pots and pot-handling gear. A sudden gust and a ship roll push her toward the starboard rail. The starboard rail is lower than the port rail to facilitate pot handling, and Tica has not yet hooked up her safety harness.

Insh reacts instinctively. Her left arm shoots out and grabs the girl's float coat while her right grabs a post attached to the superstructure. Tica struggles to regain her footing. Once they both regain their balance Insh turns Tica so she can look into her eyes. "Are you okay?"

"I think so," Tica says, breathlessly. "Thank you."

"I'm glad that happened, Tica," Marnie says. "It's a good reminder about paying attention to your safety gear. You have to stay connected to the boat in rough weather whenever possible. Here's something else, probably more important: you have to remember to think less about the wave that just hit the boat and caused it to roll and the pot to swing and more about the rogue wave that is probably coming right behind it. Cap or whoever is at the wheel tries to anticipate rogue waves, but she can't see them all."

"I'm sorry, Marnie," Tica says. "My tripping wasn't the best way to start this."

"Don't worry about it. Just focus on learning how to do your job better than anybody else. Make yourself indispensable, and all the threats you can imagine will resolve themselves into occasional irritants. Trust me on this."

Insh speaks up. "Marnie doesn't want to talk about it and I don't mean to scare you, sweetheart, but she was running the hoist on a set once and one of the clutches exploded and sprayed metal all over the deck. One of the deck crew took a piece of it in the arm, and the Coast Guard had to lift him off the deck by helicopter and take him back to Dutch. Nobody else, including Marnie, got a scratch, but she has never forgotten that day. You need to stay very alert and aware of everything going on around you while the hoist is running. At the same time you should stay behind something sturdy. If the hoist goes bad you might be the only one of us close enough to help Marnie fix whatever might be going wrong, including a serious injury to her or some other member of the deck crew."

"Thanks, Insh," Marnie says. "Those are good words for all of us. On the first five or six sets I do not want to try to set any speed records. Insh will work with me on my side of the pot when it is on the launcher. You work the other side of the pot, Tica. Remember that you have to also clear the magnetic pad, bait the pot, and throw the buoy lines into the water after launch. That means you will be moving back and forth on the deck a lot. Always, always keep your head on a swivel and your safety harness hooked whenever possible.

"You have to watch your feet and you have to always look around for things coming through the air. In rough seas things break, and things tied down can come untied. Keep your head turning and stay loose on your feet! I want to take all the time we need to get all the procedures and bugs worked out and reasonably bullet proof between the three of us for each set.

"When we get good enough that I am not worried about losing an expensive pot full of bait, or losing one of you guys over the side, we can try to speed things up, but don't count on it, and Tica, don't count on me telling you to switch sides with Insh between pot launches so you can get some experience in her position. I'll only do that if I truly think you are ready."

Insh looks toward Tica. "This is the time and place, sweetie pie, where we both snap to and say in a loud confident voice: 'Got it, boss. Let's get to work!'"

Tica smiles. In a loud and quavering voice, she says, "Got it, boss. Let's get to work!"

Marnie gives her a good strong squeeze. At a sign from Delaney, Ann Ruggles, at the helm, nudges the twin throttles forward. The Aleutian Challenger begins its four-hour run toward the southwest and, hopefully, toward enough crab to fill the vacant spaces in the live tanks below decks.

***

Dawn is breaking as Tica runs her rubber-gloved hand over the stacked ropes. She has already checked the mesh bags of bait, mostly guts and fish chunks called chum and made sure they are loose and ready to put into the pots. Marnie is a few feet away running some analytics on the hydraulic power block and hoist while talking to Sadie in the engine room. She then moves the crane boom and electromagnetic pad over to a pot on top of one of the stacks.

She synchronizes the crane with the roll of the boat to allow the swinging pad to settle on one of the pots then waits for the signal from Ann in the wheelhouse to move the pot over to the launcher next to the star-board rail. The Challenger is still underway, though, and Tica knows that there will be no pots flying around until the boat arrives and stops at the designated longitude and latitude for this first pot set.

She notes that the boat's engines, though still running, are suddenly much quieter. The Challenger is rocking slowly in the five-foot waves. Marnie looks toward her on the other side of the pot and yells out:

"Ready?!!" Tica, her right thumb held high, nods yes. Marnie raises her right hand and looks toward Insh, ready to help wrestle the pot onto the launcher.

Marnie moves the pot forward and down toward the launcher as Insh leans over with a hooked pole to grab it. Once it's on the launcher, Tica runs back to her station to get a bag of chum. She kicks herself, realizing she should have brought the bag to the launcher the first time instead of making a second trip. She resolves to do better next time.

Marnie points at the stacked ropes on the deck to indicate that she is ready to launch the pot. Tica will need to start throwing the stacks of rope after the pot as it sinks quickly beneath the waves. Then Marnie points to the buoy next to the port rail. Tica had forgotten that the buoy needs to be ready to hook onto the end of the last shot line when Marnie signals that the pot has hit bottom. Again, she resolves to do better next time.

When Marnie turns back to the hoist and launcher for one last check, both Insh and Tica put their fists up. Satisfied, Marnie punches the control on the hoist that will flip the pot off the launcher, then, once the pot was clear of the rail she punches another button to release the pot into free fall toward the sea bottom.

The large friction spool on the hoist spins rapidly. Tica grabs the stacks of buoy rope, moving to the starboard rail and throwing them over the side to follow the sinking pot as fast as she can. After a couple of minutes Marnie signals that the pot is on the bottom and Tica needs to hook the buoy to the line and get it over the side.

Once the buoy floats away, Marnie signals Ann Ruggles, who has been watching from the back of the wheelhouse, that they should move to the next set location. The entire process has taken less than five minutes.

Both Marnie and Insh come over to Tica and exchange high fives and hugs with her.

"Way to go, kiddo. You'll do fine at this," Marnie says.

Tica resolves to keep up with the fast but steady cycle of pot movements; to keep paying attention to her job and the things she can improve on, and to keep correcting herself when she knows she has screwed up. She will have time for rest and reflection once the twenty pots on Captain Delaney's list have all been set.

The remaining sets proceed without incident in the relatively calm seas. Marnie tells Insh to switch stations with Tica for the last three. The girl feels a sense of pride in her hard work, her ability to keep up and to work in sync with her two deck mates. She will eat well and sleep well tonight. She will be ready to work hard again tomorrow, even though each of the pots will be—she allows herself to hope—heavy with the money shot; lots of mature, legal crab. The only thing lingering in the back of her mind is the captain's concern about the coming storm, and the need to complete their work and get back to St. Paul ahead of it.

***

Tica is deep in dream sleep when she is jerked awake by Delaney's voice over the PA. The seas are heavier than when she went to bed. The boat is rocking from side to side and banging as it occasionally drops off a wave top.

"The storm is coming our way faster than I expected," Delaney says from the bridge. "According to the reports I'm hearing it is also stronger than I expected. It is three a.m., but I want you all out of bed, in your gear, and ready for work in twenty minutes. Sadie, I need you in the engine room as soon as possible, so breakfast will have to wait. Ann will help you keep track of things down there. I've got a pot of coffee on. Everybody start moving, now!"

"Shit," Sadie says. "You'd better move your ass, Tica. Cap gets pissed if people move too slow when there's a storm ahead and pots to pull."

"I'm up, Sadie. I'll be on the deck in a minute." A shiver of fear goes through her body as the boat lurches from the waves. She knows this will be a different experience than setting the pots yesterday.

Sadie slams the door on her way out.

As she gets out of the upper bunk, Tica notices that Sadie had dropped her Bible. She puts it in her pocket for safekeeping and plans to return it to Sadie as soon as she sees her on deck.

When she gets to the deck, though, Tica is almost blinded by the powerful lights that cast the rolling deck into virtual daylight. Marnie is already checking the power block and hoist rigging. Tica helps Marnie and Insh set up the sorting table, which will make it easier to separate the legal crabs from the immature crabs and other bycatch specified by law to be pulled out of the pots and thrown back.

They are about ten minutes from the first buoy. Marnie is at the crane, testing the boom and electromagnet and checking for problems in moving the crane arm around.

In the dark with these heavier seas this will be a more dangerous operation than the previous day. Wrestling a full crab pot from the water onto the launcher on deck in rolling seas is the most dangerous of all that they will have to do, and they must do it twenty times to recover everything.

The only thing that can make it worse is ice on the deck. It is cold this morning but, so far, still above freezing.

Insh stands near the storage space door in case Marnie has problems communicating with Delaney over the intercom. If Marnie looks at her and points up the stairs, Insh knows to run up to the wheelhouse and help Delaney get close enough to the buoy for Marnie to throw the hook to snag the buoy rope.

Marnie will then put her back into pulling the snagged buoy toward the side and then the rope itself onto the friction wheel on the hoist to begin lifting the pot. Once high enough, ideally, the pot can be tipped into the launcher. It is hardly ever that simple.

Luckily on this first recovery the pot tips straight into the launcher, and Tica and Insh are able to quickly secure it. The two women open the in-board side of the pot and start pulling the sea life out. Legal crab go into a three-foot hole in the deck that opens into the tanks of circulating sea water below decks. Illegal crab and bycatch go into a separate chute that leads off the boat back into the ocean.

As soon as they clear the catch, Insh quickly closes and secures the pot door. She signals to Marnie to lift the pot off the launcher and put it back on one of the stacks.

The next eight recoveries are equally as intense but otherwise unremarkable. The catch is still meager, between twenty and eighty mature crabs per pot. Tica is pleased with her performance, but she can feel herself getting tired.

"Can we pause for a couple of minutes?" she asks Marnie. "I need to get some food in my gut."

Marnie speaks to Delany over the intercom. "Can we pause for a couple of minutes? Tica needs some food. I'm sure the rest of us feel the same. It could turn into a safety issue if we don't get something to eat."

"Sadie," Delaney says, turning this into an all-points bulletin, "if you are listening tell me how things are in the engine room. I may need you to make us some sandwiches, but I don't want to stop running. Can you leave Ann in charge down there?"

"Things are good down here, cap," Ann says. "We could all use some food. Sadie is on her way up to the galley now."

"Marnie, can we continue running to pick up one more pot while Sadie fixes some food?" Delaney asks.

"I think so, but I think we will have to stop to eat after the next pot. We're dragging here."

"Got it. The next buoy is coming up; twenty feet off the starboard side."

Marnie snags the buoy and starts pulling it toward the Challenger. As it comes out of the water, Tica moves toward the rail to be ready to snag the pot and help Insh pull it toward the launcher.

At that moment, Tica slips on a puddle of water where ice is beginning to form. She goes down. Her head bangs into the rail on the other side of the launcher. Insh feels a bolt of fear that she might have gone over the side. She pushes herself away from the rail in order to check on Tica. There is a loud bang as the full pot slams into the side of the boat, catching Insh's hand on the rail.

Insh screams in pain. Marnie jerks around to see Insh on the deck holding her left arm, her face in a grimace. The pot, full of crab, is swinging freely and banging into the side of the boat.

Marnie yells into her phone: "Marian! Get down here now! Insh is down, and I got a loose pot banging into the boat. I can't see Tica."

Delaney bursts out of the storage space door and runs to the other side of the launcher. Tica is sitting up on the deck, dizzily holding her head with one hand and bracing herself against the rail with the other. Her head is bleeding, but she seems otherwise okay. Delaney pulls a pressure bandage out of her pocket and tells Tica to hold it tight against her head to stop the bleeding.

"Are you okay, Tica? Look at my finger." Delaney moves her finger from side to side, and Tica's eyes follow.

"I think I'm okay, cap. I'm sorry I slipped and fell. Did we lose the crab out of the pot?"

"Not yet, but we might. Can you get up? Insh is hurt. You may have to help us pull the pot onto the launcher."

Tica struggles against the rolling deck to get up.

"That's okay, Tica," Delaney says. "Stay down on the deck and hook your harness to something. Move out of the way while we get this pot under control!"

"Okay, cap," she replies meekly. Tica crawls on her belly away from the launcher toward the pot stacks. Delaney has already gone back to help Marnie tend to Insh.

Insh is still in pain, though she is sitting up. The pot is still banging into the side of the boat. It contains a nice load of crab, but it will all be lost if the banging around causes the latches on the pot's doors to fail. Marnie has tied Insh's arm to her chest until they can figure out what else needs to be done to restore a measure of functionality because they will need her active help in recovering pots. With Delaney able to help Insh, Marnie moves back to the crane controls to try and figure a way to get the pot into the launcher.

"Do you feel up to operating the power block, Insh?" Delaney asks.

Insh responds through clenched teeth and pain. "I can try, cap. Give me the control and I'll see if I can work it with one hand."

"I'm going to go to the other side of the launcher, Marnie. Let's see if we can get the pot lined up with the launcher, then give a signal to Insh to let it drop," the captain says.

"I think I have it worked out. Go ahead to the other side. I'll talk to Insh so we can coordinate the moves."

Delaney grabs a hook with a rope attached then ties the rope to the launcher. She figures this will help her snag the pot to stop the swinging.

Meanwhile, Tica has made her way back to the launcher. She does not feel confident enough to try standing, but she knows that getting the pot onto the launcher and properly secured will take at least three people.

She resolves to be one of the three, no matter what kind of effort or personal risk might be required.

While Delaney attempts to capture the pot, Tica uses her body harness to tie herself to the launcher in such a way that she will be secure and have enough slack to grab the pot and tie it down. She uses a knot that she learned from one of the whalers in Point Hope when she was a little girl.

With a lucky toss of the hook Delaney manages to snag the pot and pull it to the side of the boat. Marnie waits for an incoming wave to help her pull the pot toward the boat without it again banging into the side. She signals to Insh to lift the pot high enough that it can be tipped over onto the launcher.

As soon as the pot is down, Tica pulls her rope through the side and ties it off to the launcher to hold it in place. The pot seems to be secure. Delaney signals to open the access door on the pot, and the three women begin pulling the crab onto the sorting table.

When the pot is empty and restacked Delaney tells them to meet her in the galley so they can treat the wounds and decide if there is some way to recover the remaining pots.

Ann remains at the wheel monitoring the autopilot while waiting for a decision on where to go next. The rest of the crew sits around the crew table in the galley discussing the issue. With the help of Delaney and Marnie Tica finally recovered her footing. Marnie cleaned her cut and put a fresh bandage on it. “Do you still feel dizzy, Tica?” she asks.

“No. I am okay, I think. I’ll get myself checked for concussion at the clinic in St. Paul.”

“Insh? How are you doing?” Marnie asks.

"If your real question is whether or not I can handle the power block controls like I did on that last pot then I think I am okay to keep going until we have everything recovered."

"What about your arm?" Sadie says. "It looks pretty rough to me."

"Yeah. It hurts like hell. But the damage was probably no worse than trying to force one of your sandwiches down."

Sadie jumps up, her face hostile. "Hey, cap, I don't deserve that shit. You assholes were lucky to get anything to eat with the boat rolling like it was while I was trying to fix the damn sandwiches."

Delaney jumps up. "Hey! Both of you guys shut up! Insh, cool it. You too, Sadie."

"I'm sorry, Sadie. I really do apologize. I was just trying to break the tension, but I wasn't thinkin' right."

"That's a dumbass way to break tension, Insh," Delaney says. "Never mind. There are still eleven pots out there. You say you are okay with working the power block like you did last time. Marnie, what do you think?"

"I'm okay with Insh working the crane if she thinks she is up for it. I think Tica showed good sense when she positioned herself to help us get the pot in the launcher even though she was hurt. If she feels okay about it, I'd like her to keep doing it if we decide to recover the rest of the pots. I'd certainly like to have you on the deck with us, Marian. I need another set of good eyes, good ears, and good hands down there."

Before Delaney can respond Marnie continues. "Ann should stay on the wheel, and Sadie is going to have to be on the move between the engine room and the wheelhouse, but mostly she needs to be ready, wherever she is, to jump in to help us wrestle pots on deck."

She turns to the crew. "Is everybody cool with this? Unless there is some major objection, I think we need to get back to work on recovery. I don't think this storm is going to cut us any slack, and the sooner I can feel like we are headed to port with a load of money fish the better I will like it.

"Sadie, go up and tell Ann what we've decided, then stay with her until we snag the next pot. If you have time do a quick check on the engine room to make sure there are no flames or smoke, then stop and listen for any abnormal banging.

"Marian, go down to the engine room now while Sadie goes up to the wheelhouse. Make sure there are no new problems down there. Then come back up and help us get the next pot into the launcher. After all this I just hope that fucking pot is bulging with legal crab."

Delaney takes note of Marnie's new exercise of authority but does not try to reassert her own. She is glad to see Marnie running the boat and plans to make favorable note of it in the log, and in her next report on the trip.

***

The stormy weather does not abate as Ann Ruggles moves the Challenger close to the buoy connected to the next pot.

The deck crew is now moving more deliberately and carefully. They keep a watchful eye on Insh—so watchful that every time Insh catches one of them looking her way she flips them the bird and sticks her tongue out.

After the next pot and the next and the next are safely relieved of their loads of fresh crab, and the pots are back on their stacks, Marnie calls for a break to check on everybody and make sure there are no problems building up among the now overstressed crew.

They progress down to the kitchen, where Sadie has set up a news cast on the satellite radio. "The Pentagon has ordered two Lockheed SR – 71 Blackbird very high altitude reconnaissance aircraft aloft out of Eielson Air Force Base 30 miles east of

Fairbanks. The two aircraft are to start pa-trolling south along the Russian Coast," the radio squawks. "This is being described by Pentagon officials as a precautionary measure and should not be a cause for alarm."

"First the storm and now this." Marnie gnaws at her lip. "It's getting bad out there, huh?"

"I guess we'll see," Insh says. "I'm not worried yet, but the sooner we can dock, the happier I'll be."

Marnie takes note of Sadie. As Sadie circulates up and down the stairs between the engine spaces and the wheelhouse, her face seems to hold a persistent scowl. "What's the problem?" she asks.

"I don't have a problem, Marnie. Just let me do my job and I'll be fine."

Marnie ignores this bluster. "C'mon, Sadie. Tell me what's going on. Pulling up these last pots is going to be tough. I can't have an important member of the crew having some kind of problem eating at her."

Sadie's eyes fall to the floor. A deep grimace crosses her face like she has just bitten into the hottest of Thai peppers. Slowly she raises her face to look Marnie in the eye. "I think your little brown tattooed bitch stole my Bible."

Marnie backs away from Sadie's intense scrutiny. "What?!" she exclaims. "I can't believe you'd say that. Tica just doesn't strike me as some kind of a petty thief. Why do you say that?"

"When we got up this morning, I was the first one out of our bunk space. She was just getting out of bed. After a while I realized I had for-gotten my Bible. I went back to get it. The bitch was gone, and so was my Bible."

"Did you look for her to ask her about the Bible?" Marnie asks.
"No. I can't find her."

All right. We'll go look for her now. I'm sure this is some kind of a mis-understanding."

Just then Tica approaches the two women. "Oh hi, Marnie. Hi Sadie. I've got your Bible. You dropped it on the way out of our space this morning. Here." She hands the Bible to Sadie.

Sadie does not say thank you. After a moment, Marnie is forced to ask Sadie if she is okay now that she has found her Bible. Instead of responding to either of the women, Sadie walks away toward the engine room stairs, her head bowed.

"What was that all about?" Tica asks.

"I'd rather not say just yet," Marnie says grimly. "We do have a big problem with the recovery if anyone besides Insh and you are not able to carry their full share of the workload. I need to talk to Marian."

"What should I do?"

"I'm not sure what is going on, but for now, you should probably avoid Sadie." Marnie changes tack. "We're still rolling a lot. Maybe you should check the storage spaces to make sure the emergency gear is secure. Make sure nothing has come loose."

With that, she goes up to the wheelhouse, where Ann is at the helm and Delaney is checking the weather. "Sadie seems to have some kind of a problem with Tica," Marnie says. "I'm not sure of her ability to do her job if we try to recover the remaining pots."

"Can you tell me more about the problem?" Delaney asks. "Are drugs or booze involved?"

"I don't know. I didn't smell any booze, and she seems to be lucid enough. All I know is that she made a crude remark about Tica that sounded a lot like some kind of deeply felt racism and hostility."

"Where is she now?" Ann asks.

"She went down to the engine room."

"Did she say anything when she went down there?"

"No."

Delaney speaks up. "Do you know something, Ann?"

"Maybe. If you are okay with it, cap, I can go down and talk to her."

"Go ahead. Try to be quick. We need to decide whether we have the ability to recover the remaining pots. If you are able to sort out Sadie's problem, and you think she is ready to do her job, call me on the private line to the wheelhouse. I will need to talk to her to judge for myself her state of mind. Marnie, take the wheel while Ann and I try to work this out."

After Ann leaves, Delaney speaks again. "Okay, Marnie. Now tell me what Sadie said that gave you such concern."

"Sure. She thought Tica had stolen her Bible. She called Tica our 'little tattooed brown bitch.'"

"'Our'? She said 'our'? What the fuck is that about? Jesus Christ. How did we miss this?"

"I don't know, Marian. Maybe all the stress of Insh and Tica getting injured combined with this extended storm condition got to her. She didn't have to deal with this much stress on her two earlier trips with us."

"Let's assume that Sadie regains her senses. What do you think about continuing the recovery?"

"I'm actually more concerned about Insh," Marnie admits. "She is putting her heart into it, but I can tell she is in pain. Fatigue will start creeping into her reaction times. Assuming Sadie gets herself under control let's recover five more pots and reassess. Let's assume for now that Sadie is not available to help me on deck with pots. She will need to stay in the engine room. After recovery I think you or I may need to swap bunks with Tica, and we should otherwise try to keep her away from Sadie."

The phone rings then; it's Ann with news. "The captain should come down to the engine room. Sadie wants to apologize for messing things up."

"I'll be right down," Delaney says. "Don't go any place, Ann. I'll have some new crew assignments."

***

After their conference, Delaney is satisfied that Sadie has calmed down enough to work. As Marnie suggested they stopped after recovering five more pots. Insh is in serious pain but will not let Marnie call a halt to any further recovery. The four remaining pots are collected, and Delaney takes the wheel, turns the Challenger toward Saint Paul, and then sends everybody except Ann Ruggles down to their bunks to get some rack.

Delaney briefly considers telling Tica that she will need to stay on board rather than get off in St. Paul, but she ultimately rejects the idea. Some time apart from the stress of crew life may be good for the girl.

The boat rocks and rolls on the return to St. Paul, but the trip is otherwise uneventful.

PART 2

The seas are still rough near St. Paul Island, but the skies are clear. Delaney looks around for the best approach to the breakwater and the docks where they can unload their catch. She will also need to take on additional fuel for the return to Dutch Harbor. Marnie is working the deck with the other crew during the docking procedure, and Tica is maintaining a watch on the radio and radar to keep track of the various ships and their movements.

Since there are a fair number of boats moving around the breakwater and docks, Insh remains near the engines. She listens carefully through headphones for Tica's monotonous repetition of the phrase: "All okay on deck."

Delaney massages the twin throttles in a way that eventually causes the boat's bumpers to lightly touch the side of the dock. Marnie jumps ashore to catch and secure the forward line and then the back line.

Finally, Delaney announces that the boat is docked and secured. She shuts down the engines and grabs her briefcase full of paperwork for the Fish and Game Department's fish count rep. Tica asks if she can use the satellite phone for a call to Kotzebue. Delaney waves her hand in approval.

Tica steps to the back of the wheelhouse to use the battery-operated sat phone. She decides to try to call her friend with the Cessna 208 Caravan, Benny Hanford. There is no answer at his number, so she tries another friend, David Washington, a son of one of the "white Eskimo" families who came to the Kobuk Valley in the sixties to live off the land.

Most of the white Eskimos have now moved on to other things. One of the children of those families, for example, has now grown up and written two best-selling books about his time up on the Kobuk, learning how to make his own hunting gear and sleds and how to make his own clothes out of wolf and caribou skins. Eventually he moved to Anchorage to try to learn how to live in the modern world.

***

Tica waits to see if David picks up. She tries to contain her excitement when she hears: "Hello?"

"Hi David. This is Tica. I'm calling from St. Paul, and I am trying to find Benny. Do you know where he is?"

"Jesus, Tica. Your mom has been wanting to know where you are. This war talk is scaring her."

"I know, David. I'm scared, too. I'm trying to get home, but the only way I can get there today, or tomorrow is if Benny will agree to fly down to St. Paul to pick me up. Do you think he will be willing to do that?"

"I don't know, Tica. Flying is getting to be more of a problem every day because of the war talk. I'm not sure the feds will let anybody take a civilian airplane into the air from Kotzebue. I think I saw Benny on the street a day or so ago, so I'll go look for him. How can I contact you?"

"I'm calling from a sat phone, but you should be able to get a message to me by marine radio. I'll stay on the boat until I hear from you. If you are going to look for Benny, I'm going to see if I can get something besides fish to eat somewhere in town. I need to get off the boat for a while. It is one p.m. now. I'll be back here by three p.m. Find Benny first, but then please go tell my mom that I am trying to get to her as fast as I can."

"I'll tell her. Don't worry. I'll call you by three and let you know what I find out about Benny."

"If Benny offers to take off right now and head straight to St. Paul, please don't discourage him. Okay?"

"Got it," David says before hanging up.

Tica hangs up, too, then walks off the boat and heads up the dock toward the small fishing village. She hopes to find a burger and beer and a quiet place to sit and think. She is reminded of her college days in the snack shop, and of Ernest Hemingway in one of her literature courses: 'a clean well-lighted place' would work best for her. She does not want to discuss her concerns about Sadie until they both have had some time to think it through. She wishes her mom was with her now to talk.

Homesick, she thinks again of the aunties who gave her the beautiful tattoos. She wishes she could feel again their soft touch and soothing words as they sang to the many animistic spirits of the Inupiaq to grant Tica the strength she would need to be in this world.

When at home in Point Hope, Tica has many opportunities to talk among family members and friends when important problems need resolution. Not all problems can be resolved, though, and Tica's problem with Sadie seems to be one of those. There are no family or friends to talk with in St. Paul.

She will have to be like the white people and work it out on her own.

***

Sadie has never been able to look Tica in the eye and has almost never engaged her in conversation. When she has offered to help with cooking or even dishwashing in the galley, Sadie has been dismissive.

Tica thinks about Delaney's rule that every crew member either get along, get her problems out in the open, or get off the boat at the next port. Without her emphasis on this simple and basic rule, Sadie would be nothing more than a very rude person to Tica, and perhaps to all the others. As such, Tica knows it is up to her to try to calm whatever problem is roiling Sadie's brain.

Tica finds no beer and no burger, but the Filipino food served up at the Trident Café, a short walk from the dock, is tasty, varied, and plentiful. She decides to eat quickly, then take a short walk around the town, possibly up by the Russian Orthodox church. She will sit in the church for a while in hopes of calming her nerves before she has to talk to David again.

If David can't find Benny, she won't be able to fly north that day. She will have to decide whether she needs to get back on the boat and go back to Dutch with the crew or catch the daily Alaska Airlines flight out of St. Paul to Anchorage the next day.

That makes it a two- or three-day trip through Anchorage to get back to Kotzebue. Once in Kotzebue, she will still have to find a ride up the Kobuk to Ambler.

With all of this war business making it difficult to know if planes will be grounded or missiles launched, Tica needs to avoid spending all that extra time getting to Ambler. She hopes that David will be able to make contact with Benny, and that Benny can come down this afternoon to pick her up. If she can get back up to Ambler that night, it will make everything much easier for her mom and herself.

***

The call comes in at three on the dot.

"Tica. This is David. I got hold of Benny. You're in luck. He has been discussing a charter to fly down to St. Paul and St. George to pick up some of the field guys working with the utility company down there. He will bring everybody back. The military is involved, and it looks like they will approve the flight because they want those technical skills available in Kotzebue while this war threat is going on. He's got room for you if you can travel light and be ready on the runway with the other two St. Paul guys when he comes in to pick you up."

On hearing David's words Tica finds herself shaking with anticipation. She had been sitting in the wheelhouse with Delaney, Insh, and Marnie when the call came in. "I'll make sure I'm ready and standing at whatever time and in whichever place Benny wants me to stand."

She looks at Marnie, whose eyes were downcast. "I hate to ask, David, but one of my shipmates wants to come with me to Kotz. Do you think there is any chance that Benny would allow that?"

"I don't know, Tica. It sounds like he may have a full passenger load already. I think I can still talk to him by cell phone. I'll ask him to set a course for St. George first, then come back up to St. Paul. You should be able to contact him with your own cell phone in about three hours, so you can ask him yourself. I know he will have to top off his fuel tanks for the run to Kotz, so you should plan to meet him at the fueling pumps at the south end of the airfield.

"If it's possible to shift stuff around in order to fit the both of you onto the plane that will be the place to do it. Once again, plan to travel with nothing more than a toothbrush. Some of those tech guys may have specialized tools that will have to stay on the plane. I know you can deal with it, but you might both have to sit on the laps of some random guys for several hours during flight. Can your friend handle that?"

Marnie looks Tica in the eye and nods yes. "She can handle it," Tica says.

"Okay. I'll try to connect with Benny about his route and about your traveling friend. If there are any initial hang-ups from him, I'll call you again on the sat phone. Make sure your phone is charged up. It will also be good to have a charged spare handy. You may have to do some negotiating with Benny and, probably, with some of the passengers to make this all work."

"Thank you for all your help, David. I'll call you on the sat phone as soon as Benny and I have worked out all the details. One of us will contact you from the air as soon as we can connect with ground communication in Kotz. Thanks again. You've been a huge help. I owe you so much."

***

The sun has been below the distant horizon for a few hours, leaving a fading twilight glow by the time Benny is able to contact Tica.

"I'm about an hour out of St. George," he says. "David passed your message along to me. I will do everything I can to get you and your friend back to Kotzebue tonight, but it may require some work and some dis-comfort on your part. At this point I don't know who my passengers will be, or how many there are. The military is paying for this flight, and they are handling who sits in which seats and how many pounds of freight I will be carrying.

"They know what you want to do, and they said they will do everything they can on their end to make it work. I'm going to ring off now, though, so I can concentrate on St. George. As soon as I'm airborne out of there, I'll contact you again. Make sure your cell phone batteries are charged. I'm glad you are okay."

"David says I can't take anything more than a toothbrush," Tica says. "Is that still true?"

"Be prepared to travel with as little as possible. You can bring some kind of a travel bag with you to the airport, but if I say you have to leave it on the ground, you can't argue with me. Okay? I'm sorry to be such a hard ass."

"It's okay, Benny. I understand. I'll talk to you soon."

"One more thing, Tica. If I say you can go, but your friend has to stay on the ground, you can't argue with me about that either. Am I clear?"

"Yes, Benny. That is very clear. I'll make sure she understands. Her name is Marnie, by the way." Tica closes the cell phone link. The entire crew of the Challenger is now gathered in the wheelhouse.

"Did you get that, Marnie?" Tica asks.

"Yes. Thanks for doing all you can to help me."

Tica turns to Captain Delaney. "Well, cap, in a way whether Marnie comes with me or goes back to Dutch with you depends on when you want to get underway. Do you know yet?"

"I can wait for a couple of hours. I think the Siberian weather is still moving in our direction and is likely to be moving in tomorrow, so I will want to get underway before midnight tonight. You and Benny will need to think about beating that storm, too. Stay alert, and keep things moving."

"Marnie and I need to figure out what we are doing, so we'll come back up as soon as we are done in an hour or so. Then it will be a matter of waiting for Benny's call. Can one of you guys see if you can line us up with a ride to the airport... and back if necessary?"

Insh gets up to get a cup of coffee. "I'll make sure you have transportation, sweetie. Don't worry."

"One more thing," Delaney says. "Tica, will you come out to the deck with me?"

"Of course."

Outside Delaney turns to Tica. "Have you talked to Sadie within the past several hours?"

"Yes," Tica says. "I sought her out. I told her there was some tension between the two of us, and I felt I needed to try and resolve it. We talked for a few minutes."

"Is anything settled?"

"I don't know, cap. I told her I would like to resolve whatever problem there was between us. She did not really respond to me. Finally, I offered her a decorated talisman from my village that I thought might please her. She accepted the token but showed no appreciation or gratitude before she left me."

"I'll talk to her."

***

Benny calls Tica about two hours later, around eleven-thirty. "I'll be on the ground in about thirty minutes. Can you guys meet me at the fuel pumps at the airport?"

"Yes. How much space do you have?"

"I've got one seat with me in the cockpit, and room under it for one travel bag. There are ten passenger seats behind the cockpit. The other passengers are all men of various ages. One of you may have to sit on the other's lap, but I can't have you doing that in the cockpit. I may have to move one of the guys into the cockpit seat, and you will have to share his seat in the back. Will that work for you?"

"We'll make it work," Tica replies.

"Good. I need to warn you that we have weather coming in from the west. It's already pretty bumpy. I've got one guy freaking out in the back. I've warned him that the weather may get worse as we get closer to Kotzebue, but If he wants to stay on the plane, you and your friend—Marnie, is that her name? —may have to stay strapped in a single seat."

***

The crew on the boat assure Tica and Marnie that they will have bunks on the Challenger whenever they want to return to work. Hugs and kisses, even a few tears, are shared all around, and the women are on their way. The airport is about two and half miles from the dock. Insh takes them in a jeep she borrowed from a security guard at the fish processing plant. He had been walking the dock on swing shift.

"Sadie gave me some cookies to help convince him that loaning his jeep to me was a very human thing to do. That plus showing some decolletage had the effect I was looking for." Insh gives the travelers a lewd wink and a big smile as she pulls the jeep away from the Challenger. She stops, though, as Sadie walks in front of the jeep.

"Do you need something, Sadie?" Insh asks.

"I have something for Tica." She approaches Tica's side of the jeep. "Thank you for the nice gift. I want to give you something in return."

Sadie lays the Saint Christopher medal in Tica's hand. The two women look at each other.

"Thank you, Sadie," Tica says. "I am very pleased with this."

"I hope you have a good trip," Sadie says.

Tica reaches out to give Sadie a hug, which she accepts before turning abruptly away.

"I'm sure you can't wait to get home to your mother," Marnie says, once they're standing on the airfield tarmac.

Tica smiles. "Yes. She's everything to me. I think you'll love her." She chuckles and kicks the toe of her boot into a nearby drift. "She's got this almost translucent blue/violet stone that her high school boyfriend gave to her. He wanted her to have it even though she started seeing Dad while the boyfriend was driving attack boats on the Mekong Delta in 'Nam."

Marnie's mouth drops open. "No way."

"Yes." Tica pops the word on her lips. "It's beautiful, though. I under-stand why she keeps it. We don't know where it came from, but it's like family tradition at this point. If we evac from Ambler, she'll probably take it with her."

Twenty minutes later the runway lights come on, lighting up the entire area around the airfield. A couple of minutes after that, the women can hear Benny's plane approaching for a landing to the north. They see his landing lights come on.

Westerly winds across the runway seem squirrely, but not too strong yet. Even so, the wings dip up and down in response to sudden gusts. Benny brings the plane down smoothly, first the left rear wheel, then the right, then the nose. The touchdown is followed by a reduction in the whine and a change in pitch of the turboprop engine.

Tica speaks into her cell phone while Benny taxis the plane to the pumps. "Nice landing. I am so glad you were able to get here."

"Hi, Tica. Nice to know you were waiting for me. I think you may be in luck. I have to talk to my freaked-out passenger about his options if he doesn't want to continue the flight, but I may have two seats for you and Marnie."

"If he is that worried about flying in this weather," Tica suggests, "maybe I can persuade my captain to take him on the boat when she takes it back to Dutch this evening. I'll introduce him to my shipmate Insh if you want me to offer that suggestion."

There is a pause as Benny considers this. "I'll discuss it with him. Just be ready to grab your seat, or seats, and get strapped in as soon as I finish fueling. I want to do everything I can to beat this storm. We will all talk further in a couple of minutes."

Tica turns to Marnie and Insh. "What do you guys think of me offering that guy a place on the boat when you all leave for Dutch?"

"I don't have a problem," Insh says, "but why don't you call the cap now and find out for yourself before you talk to this guy?"

"Good idea." She punches a number and waits for Delaney to pick up. "Hi, cap. It's Tica. One of the passengers on Benny's plane has gotten into a twitch over this storm and the bumpy weather on the flight up from St. George. Can I offer this guy some space on the Challenger so he can get to Dutch with you and the others? That way he can catch the next flight to Anchorage, and possibly a same-day flight from Anchorage to Kotz. If he goes with you, Marnie and I will both have a seat on Benny's plane. Otherwise it will probably be a very uncomfortable flight for both of us."

"What do Marnie and Insh think about this?" Delaney asks.

"I think Marnie will ride in the baggage compartment if necessary. Insh says she has no problem with the idea of a male passenger on the boat. "

"Have you met this guy yet?"

"No. They are still taxiing up to the pumps."

My main concern is that Bennie's passenger might be drunk or drugged up, or he might have a gun and some kind of hateful attitude toward women, or some kind of combination of all of these. I can't allow anybody like that on my boat unless he is so drunk that he will pass out and stay passed out all the way to Dutch.

"I'm sure you are in a hurry to get this resolved and get your flight underway, but you are going to have to make an assessment of this guy's condition and be ready to assure me that he won't cause me or any of my crew any problems. If he looks and sounds like he is not drunk or drugged, and he is willing to let me lock up his baggage, I'll probably be okay with him coming aboard. Are we clear on that?"

"Yes. Yes, we are, cap. Benny is just now pulling up to the pumps, so we should have an answer on this in a minute or two. I'll call you back."

Benny brings the plane up to the pumps and shuts the engine down. He jumps out of the hatch next to the pilot's seat and opens the back hatch to the passenger compartment. A guy gets out who looks a little wobbly, but otherwise alert. Benny points toward Tica and the other women. Then he grabs a stepladder to begin the fueling process. He yells to Tica: "I'm only on the ground for about ten minutes, so hurry up and get your business done."

Tica steps forward to introduce herself, Marnie, and Insh to the man. "I'm Tica. I understand you want to get off here."

"Yes. I am very much afraid of flying even in the best of circumstances. When this trip came up, I wasn't really given an option to say no. I did grab a bottle of whiskey to calm my nerves, but the bumpy flight from St. George caused me to throw it all up. The pilot said you might have a bet-ter way for me to get to Kotzebue without losing too much time getting there."

"I think so. I'm a member of the deck crew on a crab fishing boat. The boat is tied up down at the docks here, waiting for the three of us to work out a way for two of us to get to Kotz tonight on this airplane. The captain on the boat is named Marian Delaney. As soon as we are all ready, she will depart with a five-woman crew for a fast trip to Dutch Harbor. It is up to Captain Delaney to decide whether you can have some space on the boat. I can connect you with her by phone. That may be the best and fastest way to work this out."

"I am willing to do whatever the captain asks in exchange for a ride with you guys."

"Do you have any booze or drugs in your luggage?" Insh asks. "Any guns?"

"No booze or drugs. I'm the biggest straight arrow you've ever seen except when I fly or have to think about a departing flight that I have to be on. I do have a pistol, a .38-caliber revolver—it is Alaska, after all—but I will happily give it over to the Captain during the trip if she allows me on."

Insh turns to Tica. "He sounds good to me. Go ahead and get the cap on the phone. I'll give her a briefing, and then she and he can work out any final details."

Tica approaches Benny on the ladder while he fuels one of the wing tanks. "I think we are going to be able to work it out with your passenger, so tell us where you want us to sit on the plane."

"Well, the main thing is that I want you to sit in the copilot's seat. It is not really a copilot's seat because there is no aileron control there unless I pass mine over to you. Didn't my other pilot, Timmy, throw the aileron control over to you once?"

Tica nods.

"Good. I'm glad to hear it. There is a seat in the back. Have Marnie take it. I want her to stay alert to the moods of the passengers. If she sees one of them getting antsy, she needs to get up and try to calm him down. Tell her everything about the flight and what I expect of her as I have told it to you."

"Got it." Tica goes back to retrieve Marnie and check on Insh and the sickly passenger. "Have you guys got it worked out with Cap?"

"Yup, sweetie pie. Everything is fine. Give me a hug and I'll be on my way with Tom, here," Insh says.

"Good luck, Tom," Tica says. "I hope you can make it to Kotz tomorrow, so your boss doesn't get pissed at you. I'm glad this all worked out."

***

The flight gets off the ground at twelve-thirty a.m. Benny puts the plane into a steady climb to its maximum operating altitude at 25,000 feet. He turns the autopilot on. There are some bumps, but nothing scary. Benny hopes to find smooth air at altitude. He flicks a button to drop the oxygen masks down from the passenger ceilings and asks Tica to make sure every passenger knows how to put the mask on. Most of the passengers seem to be sleeping. Tica has to wake them up to help them with their masks. She motions for Marnie to work her way from the back to the front to help the passengers behind the cockpit. Marnie seems grateful to have something useful to do.

As Benny turns to pick up his navigation signal Tica looks back to the west and notices the City of St. Paul like a ring of bright diamonds floating against the smothering black velvet of night.

Tica thinks she sees a flash of light far to the southwest in the direction of Attu, the westernmost of the Aleutian islands, but it comes and goes very quickly. It does not linger in the clouds, and she cannot be sure of what she saw. Shortly after, Benny and Tica both feel an abrupt bump of turbulence, but it does not last long. Benny continues to cruise at 25,000 feet, above most of the weather. There is enough of a moon to cast light and shadows on the cloud tops.

It is a beautiful night above the Arctic Circle. The Aurora Borealis is sometimes a beautiful ring when seen from an airplane at this high altitude. Tica hopes she will be lucky enough to see a display.

***

Benny is about an hour out from Kotz. He is having trouble raising anybody on the ground. He needs to hear from approach control for navigation guidance, and he needs the runway lights turned on, but there is no answer from the tower. He switches to the military channel that has been issued to him for this flight.

"Kotzebue tower, this is Cessna N393 Charlie Foxtrot coming in with military cargo from St. George and St. Paul. Over?" The silence on the other end is deafening. Benny repeats the call with more vigor. "Kotzebue tower, this is Cessna N393 Charlie Foxtrot coming in with important military cargo from St. George and St. Paul. Is anybody down there? Please respond. I will be descending for approach to the outer marker for runway two seven in about thirty minutes. Please respond. I need some-body to turn the lights on. Over?"

He turns to his copilot. "Tica. Take the radio mic and repeat the message I just gave them until somebody answers. I need to check some things in case everything is all fucked up on the ground, and I have to do some emergency procedures. Stay on the military channel, but if you are not having any luck, try some of the other channels.

"Try your friend David in Ambler on your cell phone. He may be monitoring your number. Try any number you can think of. We are running out of time and fuel, and I need to know what is happening on the ground."

Tica repeats the call to the military twice more with no success. She picks up her cell phone to make a call to David and is surprised to see that he left her a text message a few minutes before. She texts back. "David. Thank god. We are approaching Kotzebue but there is no answer on the radio, do you know what is happening?"

Tica breathes a sigh of relief as David responds with a text. "Try calling me. I think we can still talk, but I am not sure for how long."

David launches right into it when she calls. "I am going to speak fast, Tica. The military has taken control of all local communications and they have shut down the airport at Kotzebue to any civilian traffic. There was some kind of a bomb, or more likely a missile, that landed on Attu at the end of the Aleutian Chain a couple of hours ago. All we know for sure is that there is a military outpost out there that has gone totally silent.

"I'm able to talk to you because the mining operation at Bornite has its own comm link that doesn't depend on the usual civilian channels. The military's reach is long, so I don't know how long the Bornite link will last. You need to listen carefully to everything I say. You need to divert your flight to the field at Ambler. Electricity in Ambler is generated by backup diesel power locally, and we can light the runway when you get within range.

"A friend of mine in Ambler is a radio geek. We think we can put out an ADF direction signal on a regular radio frequency that your pilot can use to get through the clouds. Once he gets through the clouds, we will have the runway lit, and he can make a normal VFR landing under visual flight rules. Have you got all that?"

"I think so, David," she says, "but I want you to talk to Benny. I don't want to screw up the pilot talk. He needs to hear that directly from you."

Benny is running instrument checks, but he can see from her facial expressions and hand gestures that the phone call is important. She hands the phone over.

"Hello David. What's up?"

David repeats all that he had told Tica. Benny asks that the radio setup in Ambler play some Johnny Cash, so he will know that the ADF signal is specifically for him. "'Ring of Fire,' one of my favorites, will be the best, " he says.

"I think your best approach is runway 1 to the north," David says. "You are familiar with the holding patterns and surface problems on that run-way, especially in the dark with a crosswind, so I know you can handle it. I'll try to run the grader over the surface, but no guarantees on that. Remember that if you have to abort a landing on Runway 1 you need to bear right to avoid the hills beyond the north end of the runway.

"I'm going to head over to the radio setup now to make sure they are on air, and they have 'Ring of Fire' on the turntable. I'll keep this phone connection open as long as I can, but if I lose contact with you, I may not be able to reconnect. Keep your ADF receiver on. We may be able to get a signal up, but we may not have much range—possibly no more than twenty-five nautical miles. We can also put up a shortwave signal, but I don't know if your plane is equipped to receive it."

"My GPS has not gone down yet," Benny says, "so I still have good lat/long info coming in. I'm turning now to set a course for dead reckoning to Ambler while I still have GPS. I am showing about forty-five minutes to Ambler, by the way. We have shortwave receiver capability, but I am not sure I can get a direction out of it except by monitoring signal strength. Do everything you can to direct the beam along magnetic bearing 050/310. We will work on receiving on this end."

"Good luck, Benny. This may be the most important night of your life. Put Tica back on the line. As I come up with new information, she can relay it to you. When you get your plane down on the ground, we can all go get a beer. I'll even buy. Just don't fuck it up."

"Thanks, man. If I come in upside down, I'll try to walk away from it. If I walk away, it's a good landing, right?"

Tica relays to Marnie all that they had learned in the past several minutes. She needs Marnie's calm in a crisis to help deal with the passengers.

Tica works to be ready for whatever Benny needs to bring this difficult and dangerous flight to a successful conclusion. As she notes the instructions he gives her about the function of the various radios and navigation devices, her mind wanders to Benny's life growing up in Kotzebue.

***

Benny's father and grandfather were both whaling captains. As captains they were in charge of a walrus skin boat about as long as the fuselage of Benny's airplane. When a captain and a whaling crew of six paddlers hunt whales in choppy seas, with waves lapping above the low gunwales and into the boat, it becomes the whaling captain's job to keep everybody calm and focused on the job at hand.

Even on the deck of the Challenger, with the seas running high and the heavy pots swinging around, there is nothing to compare to the dangers of whale hunting far out to sea in an umiak, a small wooden boat covered in walrus skins.

The resolution of such frequent danger is deep in Benny's blood. Tica looks at Benny in his baseball cap and headphones; she can see him standing in a parka made of seal gut, fringed around the face with seal skin and wolverine fur. She can see him balanced in the prow of a rolling umiak, with a harpoon at the ready, scanning the horizon for the telltale venting and rising steam of the whale's exhaled breath.

Whale hunters say that the whales talk among themselves in order to deal with the threat, but you can only hear them if you are at peace with yourself in the midst of surrounding dangers. Whale sounds are not easily heard by human ears. Usually, they can only be felt as vibrations through the skin of the umiak, a feeling of a high-pitched, humming ululation.

To Tica Benny, now, is like that whaling captain.

Benny turns away from his instrument panel to show her the combined compass and ADF signal gauge. He taps the gauge with his finger and says: "I want you to keep a close eye on this instrument now. We are get-ting close enough to Ambler that we should soon be able to hear Johnny Cash playing 'Ring of Fire.' That means we can pick up the radio station ADF signal on this gauge. When that double-barred needle jumps, let me know right away."

After a few more minutes, the double-barred needle jumps. Tica touches Benny on the shoulder and points at the gauge as it jumps up and down. Benny adjusts his course to align with the bearing indicated by the ADF signal. He cuts the power, pulls the wing flaps out, and begins a fairly steep descent through the cloud cover over Ambler.

Shortly, the runway comes into view beneath the clouds. Controlling his touchdown Benny finally sets a left wheel down, but the strong crosswind causes the nose to drift to the right. Benny knows he has one chance to correct the drift before the nose drifts out of control in the sudden gust.

As soon as the nose begins to drift, he instinctively jams the left rudder pedal and pulls up on the handbrake next to his seat. In a moment the nose comes back to the left, and the right wheel comes down to kiss the runway. The nose wheel follows. All the passengers, including Tica, clap and yell their approval for this masterful job by a man who, they are now convinced, is the very best airplane pilot they have ever known.

***

David has brought Tica's mother, Mary, to the runway. Tica grabs Mary and David and hugs them both, hard. None of them can keep from breaking into tears at the reunion. Tica motions for Benny to come over and join in. He only hesitates a moment before joining this private moment among loving friends and family.

David had stashed some drinks and sandwich fixings in the bunkhouse in case anybody was hungry. It's after the local tourist season, so there is plenty of room in town. They are grateful that, in addition to still being alive, they will also have a warm place to sleep for the night.

Tomorrow will be a whole new day—possibly, a whole new life and a whole new world to live in.

Tica, Mary, and Marnie pile into David's jeep for the ride to Mary's house. She lives a few miles west of town in a log cabin her husband, Antonio, had built out of local logs. Once inside, Tica cannot help but feel a great sense of relief. She collapses into an overstuffed chair near the wood stove where David works to get a fire going.

"Your dad is fine," Mary begins. "Unless some problem comes up at the mine, he will meet us later. I have something to tell you, my sweet baby. We are going to leave tomorrow on a hike to Alatna on the Koyukuk River. We will go up the Kobuk by boat and over the hills on foot. It will take us about thirty days, I think. I have already packed everything we will need."

This revelation takes Tica aback. "Mom, why in the world do you want to do this? And why now?"

David and Marnie sit mute on a nearby couch.

"I have been thinking about this ever since we elected that man, whose name I will not say, to the presidency," she says. "Whatever terrible thing has happened at the end of the Aleutian chain early this morning, it is only the first of many things that will now come home to disturb and disrupt all the good parts of our lives."

"I don't understand, Mom. I hope you will explain whatever is going through your mind to come up with these thoughts."

"It is simple, my sweet one. There is a dark cloud among the Inuit. It came with those who live only to take our culture, the things we make that are beautiful, the things that make us the real people. They do not know the real people, and they do not know the spirits of the earth. The dark cloud will not go until those who only have values in money and their objects of vanity have gone from us. I think it is time that we go back to our relatives on the Koyukuk at Alatna until all this evil has passed. That is what our ancestors would have done, and this is what we must also do."

"We have two guests, David and Marnie." Tica gestures toward the two figures on the couch. "Are you thinking they will go with us?"

"As the real people we have an obligation to assure the safety, security, and lodging of all our guests. Of course, they may come with us, but they themselves must choose to go."

The conversation has grown intense. Tica sits quietly for a moment, distracting herself by looking around the walls of her mother's house.

There, several beautiful works hang, representing the art of the Inupiaq, the Athabaskan, and members of lesser known tribes and language groups in the valley between the Brooks Range in the north and the Alas-ka Range in the south. Mary has many caribou skin masks wearing many expressions, all made by Nunamiut artist friends who live and work in Anaktuvuk Pass.

One of the works on the wall intrigues Tica. She has not seen it before now. It is made from a variety of grasses, twigs, and dried berries, all laid on a background painted to convey an impression of a distant range of mountains shrouded by low clouds. It reminds her of the mountains of the distant, darkened Brooks Range, with bright streams zigzagging down their front face to connect in a braided outflow into the great Yukon River.

The artist has shrouded the berries with some colored lichen in a way that makes them appear, in the distance, to be clothed in bright fashions. Some of the twigs the figures hold represents weapons and packs, as though they were hunting, or just traveling, or perhaps they are on their way to war.

"Did you do this, Mom?" Tica asks.

"Yes."

"When did you do it? What does it represent?"

Mary looks away. Her lips crinkle in a small smile. "I had a vision in the spring last year. I remember my grandfather telling me about the time when a new group of white people, like those who built the bombs used to destroy the people of Hiroshima and Nagasaki in Japan, who put money and things above the good of the Inupiaq and even above the good of their own white people, came to our lands near Point Hope.

"They planned to make huge holes in the earth at the mouth of a creek the white man now calls Cape Thompson, but that we have called since time immemorial Ogotoruk Creek. We hunted the life-giving caribou there. The white men would have used their terrible bombs to create a harbor where no shipping would ever occur, bombs that would take the life-giving power of the sun to destroy life, to poison and burn everything down to the bare rock.

"These new weapons transform this power into an evil and all-consuming fire that eats every living thing. It is the true Book of Revelation. This new group wanted to blow big holes in the ground where the caribou feed. No one knew why. Apparently, it was for their own strange purposes. They gave no thought to the needs of others who lived there and who hunted the caribou and grouse there.

"My grandfather gathered his two young daughters and son up with his wife and all went to visit their relatives in Alatna then, too. That is why I want to go to Alatna now. It is because my grandfather took our family there to keep them safe from the warped and corrupted sun power that man has made and that now threatens us again from the south."

"But how will we travel?" Tica asks. "I have never been over the trail to Alatna."

"The trail is an ancient one, though not much used these days. There are ancient stories from our family about those who took the trail to escape the many threats that came up the Kobuk river valley as the glaciers melted and the seas rose. Those of our family who made the long journey on foot came to the river we now call the Koyukuk, where they established the community of Alatna. In the early days, some crossed the river and continued up to the pass in the Brooks range where they established the community of Anaktuvuk.

"Anaktuvuk is a place of great beauty. It is a magical place where the caribou in their great numbers leave their winter grounds in the mountains, headed for their traditional birthing grounds on the Arctic Coastal Plain in the spring. From Anaktuvuk we can look north toward the Arctic Ocean and south toward the Yukon River. It is a place of great power, and the Inupiaq feel the constant pull toward that place."

David speaks up. "With respect, have you been over this trail, Mary?"

"No, I have not, but my vision of the journey I must take is clear. I have no doubt that I must go with my daughter Tica, and I have no doubt that I will find my way. You and Marnie are welcome to join us, as we will need the kind of help that two strong young people can provide."

"There was no room on Benny's airplane," Marnie says, "so I couldn't bring any of my winter gear from the Challenger. I have no clothing for being outside for any length of time, let alone to start a hike of several days above the Arctic Circle in late Fall."

"I don't think clothes will be a problem for you, Marnie, nor for my daughter. I have been gathering a lot of clothing and gear for such a journey for over a year now. I assume that you have plenty of gear available, David, and I assume you will be willing to share as necessary—that is, if you decide to go."

"This is such an amazing night," David says. "I thought my work would be done here as soon as Benny got that overloaded airplane on the ground with all its precious cargo."

He turns to his fellow newcomer. "What do you think, Marnie? I have enough gear for the both of us. From what Tica has told me of your work as a deck boss, I have no doubt that you will be able to deal with whatever inconveniences arise. Do you?"

Marnie averts her eyes from the group. "This is so sudden. I have so many things I must think about, but I know there is no time to think about them. Thank you, David, for your expression of confidence in me, and thank you, Tica, for letting me come on this homecoming with you. Mary, thank you for even thinking about taking a green cheechako from a cracker family in Georgia—who doesn't even have a pair of hiking boots—with you on a journey that is so important to you and to all of the people here."

She pauses for a moment. "I am ready to go if you will take me."

David claps his hands together and stands. "Marnie and I can go over to my place to start packing gear. We will see you back here in a couple of hours."

"There is one more thing, Mom," Tica says. "I want Benny to come with us. He is a good man, a strong man from a strong Inupiaq family. Can you go find him, David? I know he won't try to get back to Kotz until he is sure the runway will be open and lit. Is he staying in the lodge with the other passengers?"

She switches to the ancient tongue of the Inupiaq: "That's okay, Mom, isn't it? I worked with Benny in the cockpit as we had to try and find the Ambler runway. We were over clouds on dead reckoning, but he never lost sight of where we were going. His vision is clear. This hike and the times ahead will be difficult. I want someone of Benny's character and strength with me."

Mary walks over to her daughter and hugs her. She responds, also in Inupiaq. "You are my God-given daughter, Tica. I love you. Of course, it's all right. Bring him with us if he will come."

Benny, when presented with Tica's proposal that he come along on the hike, does not hesitate to say yes.

***

With skillful boatmanship on the Kobuk, they are able to get upriver to a point where the ice finally becomes unpassable. Mary announces that it is time to get off the boat and prepare for the remaining ninety-mile hike east to Alatna.

"If we maintain a decent pace," she says, "we should be able to get to Alatna within fifteen days. David, you, Benny, and Marnie need to find us some fresh meat and some deadwood for a fire. I want to recheck our gear, eat a good meal, and then get a good night's sleep."

Tica had been talking to the boatman back where he had put the group ashore. She turns to the group and speaks loud enough to be heard. "Rei-no has decided to come with us, too. He wants a hand pulling his boat up out of the water, so the ice won't wreck it when it finally freezes."

"Welcome, Reino," says Mary as she walks over to lend a hand.

***

Mary is a tough and seasoned traveler, but after a few days over land on the trail, she calls for a stop for the night a little earlier than usual. The hikers pool their rations and enjoy a good meal around the cooking fire. The group sleeps in a single tent to preserve body heat. Early the next morning, Mary wakes up before dawn to start a fire. As she looks up at the sky with the early morning glow telling of the coming dawn, she looks for the ancient patterns among the stars.

She looks particularly at Cassiopeia, across Polaris from the constellation known among most north Americans as the Big Dipper. At this time of year Cassiopeia is upright in her chair and in all her glory and queenly majesty.

Mary had learned about the Greek myths in college. In Greek myth, Cassiopeia is the bold black queen to Cepheus, king of Ethiopia. Cassiopeia believes herself to be much more beautiful than the sea nymphs that surround Poseidon, the god of the waters, but Poseidon becomes offend-ed at her audacity. Eventually, he casts Cassiopeia's beautiful daughter Andromeda into the path of the sea monster, Cetus, where Poseidon hoped she would be eaten. The young warrior Perseus intervenes when he destroys the sea monster with the head of Medusa before it can eat Andromeda.

Mary takes solace in this story. Unlike Cassiopeia she has brought her daughter away from threats by unnatural people. She thinks about the group's forthcoming arrival in Alatna. Allakaket lays across the river, and Anaktuvuk Pass lays farther beyond to the northeast. Soon she will be home.

A large black raven with a white feather on its right wing lands within a few paces of Mary. Over the course of their hike, the raven has followed them. He has become known to Mary because he seems to respond to a sound like Bw'-ak. She makes the sound, and the raven, as always, responds by sidling up to her and rubbing the top of its head against Mary's extended forearm. Bw'-ak and several of his avian mates had accompanied the walkers at a distance through their journey from Ambler. Mary's confidence grows in the belief that he will be there for them for the rest of their journey.

As always Bw'-ak brings a feeling of community with the animism that gives much life and hope to Mary's world. She welcomes the raven.

Mary considers that Bw'-ak will appreciate seeing the blue stone since both the bird and the stone have brought Mary such good luck. She takes the stone out of a small pouch with beadwork that she had kept close to her breast since leaving Ambler. She lays the stone in front of Bw'-ak. Bw'-ak tilts his head to look at it from several angles of view. The raven rubs one side of its beak on the stone, then the other. Then he looks into Mary's eyes, extends his wings, and flies away.

Mary follows the raven's progress out over the snowy horizon. In that moment, Mary feels that all will be well. In a few moments, the others will wake. They will make ready and will continue on their ever-shortening journey. She will hitch her pack higher onto her shoulders and follow her daughter out into the drifts.

END

Personal Note

As with any creative work as complex as this there are many people to thank for their wide varieties of help, support and assistance. Many friends and family have either purchased or received copies and most have read, commented and passed a recommendation to read to others.

The first book, Cassiopeia's Quest was published on Amazon in June of 2018. Two reviewers have commented. One review gave me 5 Amazon stars and one gave me 4. Their comments were to the point and detailed enough to be very helpful to me. I am delighted and hope others will follow their example with their own reviews.

I owe special thanks to a woman whose support for this writing project of mine has been unfailing and steadfast since I started the work. Hetty Barthel has read and edited and commented vigorously and in depth to me about many aspects of all of the 10 stories. Our working relationship is transactional. I give her help in doing oil painting; she helps me in learning music; … especially how to renew my long ago lost knowledge of how to play the clarinet. We both like hiking.

My work with my local printing company Alaska Litho, has been very solid. Their prices are right, their design and suggested content changes effective and their print deliveries on time. I expect to continue working with them for as long as I need to have words and images printed on paper.

Juneau is known around the state of Alaska, even nationally, as a small town with an abnormally large and diverse arts community. The three parts of that metaphorical milking stool are the artists, the arts organizations that find the money and the facilities, and the vendors who help the artists sell their stuff. When I have needed help from any of the individuals in those organizations, they have given it freely and effectively. I thank them for that. It is not possible to do creative work without their sincere efforts.

Finally, there is the excellent work of my talented professional editor, Jessica Hatch of Jacksonville, Florida. One of the Amazon reviewers enjoyed reading my first book (the quote was: "I couldn't put it down." Delightful.), but strongly recommended that I re-work the collection with the help of a professional editor. Jessica is that editor. She has brought industry experience, personal intelligence and a very solid professionalism to the business of editing and supporting a writer like me in the hard work I and others of her clients try to bring to the writing of fictional stories.

I have valued her work with me and hope we will work again soon.

I wrapped the new holiday release of Cassiopeia's Quest—Revelation at Juneau's long-running Public Market over the three day weekend following Thanksgiving.

I was already thinking about my next book project. I decided it would be a complete departure from my original scope and intent in the writing of Cassiopeia's Quest. However, I still needed to keep close touch with the valuable research that went into the times, places and characters that infused each of the 10 stories with meaning. Of course, I also wanted

to extend and expand my hard-won lessons in how to write stories and deepen my skills in the writing of fiction.

So...I decided to write a novel beginning after the holidays in January, 2023.

For that purpose I needed a fresh point of view about the 10 stories I had written; the novel I proposed to write and the lack of "pop" in my always, so far, hand-built book covers.

Putting "pop" into the cover turned out to be easy. My printing company, Alaska Litho, had hired a visual arts designer—Annie Kincheloe—who had all the talent I needed to give me a cover much more in tune with national book cover design trends than I could ever have achieved on my own.

For the new POV, I looked to my membership and active participation in local fiction and playwriting workshops and various meetings sponsored by a statewide non-profit corporation, housed in Anchorage, Alaska named 49 Writers. Kristen Ritter is a senior staffer at 49 writers, and I had begun to admire her very capable management of the day to day work of the organization as well as the conduct of their frequent workshops. Kristen often participated as facilitator and workshop participant in the various writing exercises.

At some point it occurred to me that Kristen might have the kind of new POV that I was looking for, so I asked if she was available for a brief consulting project separate from 49 Writers. The work would involve reading each of my 10 stories and giving me a 2–3 page story review at the end. She agreed, and I am very happy with her work. I look forward to keeping her reviews, and her contact info, close at hand as I try to put a novel together in 2023.

Jerry Smetzer, Juneau, Alaska USA; Monday, October 24, 2022

Personal Note

Reader's Guide

The blank sheet of paper.

This writing project began as a late-in-life consideration of all that I had seen and done while traveling around the world - after college and some work to earn travel money - on a solo adventure. I am a movie-goer.

Several years later I learned, from a movie of the same name, that Australian aborigines name this kind of curiosity and exploration of new places a walkabout. A young aborigine on walkabout is transforming himself from a child into an adult. He takes a knife, a bow, some arrows, perhaps, and a few other tools and personal items into the Australian outback. There he must learn to survive, alone, for several weeks; ...possibly for several months.

In the movie, a troubled and disturbing confluence of events put an aboriginal youth on walkabout with a young white private school girl and her much younger brother. They were deep in the outback searching a damp spot on the ground in hopes of finding scarce water. They are far from any hint of civilization; far from the supportive structures of human society. They have no common language. Together the three would have to learn to survive their walkabout together in order to get back to the comforts and security of their home communities.

This cinematic walkabout story has persisted in my memory all my adult life. It has become entangled with all that I have read, studied, and practiced in mathematics, history, literature and the arts in college, in working for a living, and in helping to raise the children. A desire grew in me, over time, to capture both the grossness and grittiness of the walkabout experience with the less gritty more abstract ideas gained by reading, working, and living life in places where having a warm, dry place to sleep and warm, good food to eat each day is never in doubt.

The idea of writing a book about the aboriginal practice of walkabout in some combination with a figure like Cassiopeia – as derived from oral storytelling and Greek myth - began to form over the summer of 2015. At the time I was completing work on a series of 10 oil-on-canvas paintings. I had recruited a professional model and had begun sketching representational poses in 2013. Together we created pencil, and pen and ink sketches of a few hundred portraits and poses. Along with the poses I had also googled and gathered a set of images, portraits and references to incidents and events relevant to periods of history that have become memorable and important to me over time as cultural markers.

Examples of these markers are the visual objects within the 10 oil paintings that are the visual references for these stories.

It was out of those poses, images, and web references that the topical outline and titles for the 10 pieces of short historical fiction presented here became stories.

In thinking about the paintings each of the 10 appears at first to be a kind of collage of objects. Actually, each is closer in concept to a message notepad lying next to a telephone. Over time a telephone pad becomes a palimpsest of layers. Each layer captures and preserves our most thoughtful impulses when we take note of the important moments and information given to us. With such a pad it is only the writing most deeply felt or most immediately consequential that persists on the pad as each page, or layer, is eventually torn off by use and age.

Layering is a tool commonly used in digital drawing, painting, and motion graphics programs on computers. I use brands like Adobe "Photo-shop," and Corel "Paint" extensively, along with a variety of cheap mobile "apps" to create multi-colored, multi-layered compositions of digital imagery of the kind used as references for the 10 paintings discussed here.

Layering is also at the heart of a computer's operating system and programs. When a computer is turned on, or when an app is launched, blocks of code containing groups of algorithms are layered onto the computer's memory in a specific sequence. Taken together all the layers of code in a program are what bring the computer and supported software to life as a useful tool for a writer, or for anyone trying to express himself or herself in a creative way.

One other piece of image manipulation software should be mentioned here. When working with a live model I sometimes found it necessary to visualize a three dimensioned human form in a complex, sometimes animated pose with variable lighting and orientation to a variable point of view. The use of 3D animation software gives me the ability, to look at each frame in a completed animation. With this frame-by-frame analysis I can figure out how to render and light the pose within the two dimensions available on a canvas or computer screen. For those times I use a 3D object manipulation/wire-frame painting and animation program called "Poser."

Using these software tools together in some integrated combination makes it possible to bring all the discrete pieces of digital imagery and in-formation together into a digital composition. When I am ready to paint, this composition guides my application of paint onto canvas in much the same way that a fresco artist applies his design – also known as a cartoon and paints onto wet plaster.

Each of my completed paintings are a collection of scenes and images

representing the period of history suggested in the title of the painting. As with any project involving personal expression in the public space, early promotion is important. When completed, these 10 paintings went on display in the hallways of Juneau's public radio station, KTOO, for four months beginning in December 2015.

The Protagonist.

The central figure in each painting is a visual representation of an intel-ligent and attractive young woman. Since each painting tells a story, each young woman featured most prominently in a particular panel becomes the story's protagonist. This choice of women protagonists reflects my life-long respect and appreciation for the strong, intelligent women in the arts, and among the friends and family in my life and in my workplace experience.

I had not thought of fiction in my original concept for this project. It began and continued for a few years as a project in the visual arts as I developed the 10 20" by 26" oil on canvas paintings. At first my plan would have me writing a paragraph of non-fiction text describing some of the historic, social, cultural, and political forces at work as suggested in the historical context of the paintings. I had not realized how inadequate the descriptions would be in giving force to the descriptive narrative.

Then my daughter, Megan, a professional art historian specializing in the art of Pacific Northwest Coast Indians, said she wanted to know more about the women in each of the paintings. In paraphrase, "Who is she?" my daughter asked. "Does she have a family; why is she in these stories?"

Since that day, I have spent most of my writing time trying to answer those questions. I resolved to write pieces of short fiction that link - to the best of my ability though I am no historian - the historical context in the published record with the day-to-day living experiences of my fictional protagonist. Her daily life, of course, takes place within the fictional and non-fictional people and events that take place around her. I hope by this book that I have made at least a start toward answering my daughter's questions about each of the young women in each of the paintings.

***

The tools.

Much of my working experience since college has been in information technology. Trying to maintain a high standard of historical accuracy while writing a book of historical fiction means keeping track of the web links, research data, and other information that I would have to gather, verify, keep current, and make easy to find and retrieve.

From my day jobs I know how to set up the simple file structures, web page bookmarks and databases that are expandable, integrated and reasonably secure. As part of the original painting exercise, I had developed a reference library of several books, images, and Google search book-marks appropriate to the painting and title in each panel. I added these to my newly created system of computerized file structures, web links, and databases.

So far it has worked well. My reference database now contains over 250 items. It includes books, movies, lectures, and images. Some of these references I have studied closely, others I have glanced at in order to glean a reference detail, and some I know to have been major influences in whatever writing and writing style that I hope to bring to these stories.

The reference list does not include the Googled bookmarks of relevant pages on the internet. The rough page count of web pages referenced probably adds another 200 to the reference count.

In history, maps are important. The tools available in Google Earth are used extensively in the research for each story. Without the visual imagery in the geographic facts of the historic places, these stories in text could not carry the fictional narratives. Unfortunately, Google's copyright policy in using images built with Google Earth is so ambiguous as to make them unusable for publication. Using my other image manipulation tools and royalty-free map projections I have drawn my own maps for the book's cover and for the pages titled "Story Geography and Story People" that opened each story in my original book.

In writing Cassiopeia's Quest - Revelation, however, I am trying to keep my publishing and printing costs to the bare minimum so I can keep the book's price tag within reach of less affluent readers. Therefore, there is only one map at the beginning of the book, and it is published in gray-scale. The only color in this book is on the cover.

There is one comment I should make on the use of digital search tools on the internet. They are essential in any kind of creative work – whether fiction or non-fiction - that must be tied to the "facts" of time and place on the public record. Like words, facts are slippery beasts that must be carefully hunted, and rigorously cross-checked against multiple sources. Even then there is no absolute assurance that the "fact" will stay static and impermeable from search to search. There are many reasons for this, but most have to do with the search through digital systems where each branch in the search path depends on a binary yes/no choice. Whichever choice is made, the path through the alternative choice is no longer avail-able in a simple digital search.

Such lost search paths may contain gold mines of data, but if the cross-checking is not sufficiently rigorous, they won't be found. A computerized search mechanism that keeps all the paths and their alternatives available to the search algorithm until it is complete is not yet readily available. That algorithm probably won't be available in any meaningful way until quantum computers come into more common use over the next several years.

Finally, I have become well experienced in the use of the expensive industry standard for image manipulation, Adobe Photoshop; and for words, data and presentation Microsoft Office. Photoshop became essential in designing and assembling the hundreds of images layered into each of the compositions for the story panels. I did not usually dip my brush in oil paint until I had a digital photoshopped composition file that gave visual hints of everything I wanted to say. Such compositions should also give guidance in the process of expressing complex ideas in the 2-dimensioned visual space of a piece of canvas or on the visual matrix in an electronic digital screen.

Setting these computerized structures up and loading the data and information I had already used for the paintings took a bit of time.

However, the savings in time these data and information structures make possible when looking for specific references while composing either text or images makes the time invested in design, setup, and data loading a significant factor in any time or budget estimate.

Capturing the muse.

The success of any kind of complex writing project depends on capturing the muse when she overcomes her shyness or reluctance enough to approach the writer. When she is nearby a writer must capture the moment by writing and writing, then writing some more. Any disruption that takes away from her presence at your shoulder will break the mood, the focus, and the energy. She will fly away, and her flight will destroy the writer's ability to write until she returns.

I knew this writing project would depend on having as few disruptions from the writing as possible. The time wasted trying to find specific references that are not where you thought they were is at the top of my list of time-wasting disruptions in the process of writing creative fiction and non-fiction. Properly set up, a computer data system with a system for organizing books and other physical references can eliminate much of that time and waste but they cannot eliminate all of it.

If I learned nothing else from my years working in information technology, I did learn there is no perfect digital system. By the nature of digital systems built on Boolean yes/no choices, there are many built-in failure points. The only digital systems that can function in real-world production environments are those where the error rates that kill productivity are defined and identified as to their likely occurrence. These occurrences are reduced to some agreed upon level where the cost of fixing an error can be assessed against the cost of replacing the hardware or software part that is the source of each of the errors.

Normally this error capture and correction process is done through a well-defined and well-financed collaboration among content and function specialists; a kind of collaboration that is almost never available when an individual engages in the lonely business of artistic expression.

***

What is the story line and story arc; what are the scenarios?

The visual imagery in the 10 paintings in discussion here having been set for the context of the history and the protagonist; the impulse then moves to create an identity for the protagonist – following up on the question "...who is this woman?" This is the business of writing story lines, story arcs, or scenarios. That writing now begins. These story guides help extend each painting into the fictional characters and events that occupy that piece of history under the "lens."

The terms story lines, story arcs, and scenarios may suggest something similar about story structure, but there are some subtle and important differences; If the writer has not yet moved beyond the blank sheet of paper that sits in front of him or her, he or she may need a story line. A story line, in the form of a couple of sentences, sets the initial time, place, and principle character in the story. It should give the reader at least a glimpse of what kinds of things might happen as the story unfolds.

If the writer knows where the story starts but not where the story ends, he or she may need a story arc. In other words, from the beginning how does the story unfold? Does the protagonist take a trip, decide to stay at home, get caught in a snowstorm without a coat? How do these events lead to the final scenes in the story, and how do any conflicts or threads in the early scenes get resolved in the final ones? These are questions that are dealt with in a story arc.

Most stories are made up of several scenes, or events, or incidents that show how the characters interact in various settings and events. Each of these can be described in a scenario that describes emerging conflicts or interactions among the characters that may be introduced.

***

What is the Writer's Writing Style?

A writer knows that he or she must, sooner or later, have a recognizable writing "style." A style gives the writer identity in the mind of their readers as well to himself or herself. In the simplest possible concept, a writing style is the narrative platform created by a writer as he or she refines and adjusts it to tell particular stories. As the writer's oeuvre expands the work gets more complex. Even a beginning writer may find it pays to develop a written style sheet to use as a handy reference and reminder of the writer's preferred way to tell stories, if for no other reason than to keep track of the correct and consistent spelling of character names.

A style sheet is a combination of many elements. The style manual published by the University of Chicago in 1993 is over 900 pages. Rather than try to catalog all of the possible elements of a style before beginning this writing, I felt my time would be better spent by reading books of fiction by authors whose works I have enjoyed and whose works have been recognized for their popularity or their importance to literature over time.

I resolved to read several of the books in English that have defined the diverse and deep scope of English language literature.

Don Quixote by Miguel Cervantes was a good place to start, I thought, though I soon realized that reading anything in translation was not going to give me access to the actual writing style of the writer because I was actually reading the translator's style. From then on, I concentrated my readings on writers who write primarily in English; with the exception, of course, of Ficciones by Borges.

My readings continued through many works. Middlemarch, by George Elliot, a woman, took me over 800 pages of dense text deep into the politics and social and political nitty-gritty of a small English town in the early 19th century. Another English woman writer, Charlotte Bronte, captured my attention with Jane Eyre, a much more readable book made even more accessible by several movies by different directors and actors.

Writers known for their extreme styles also got their share of my attention. James Joyce's Ulysses had sat on my shelf, unread, for years. I finally resolved to read it, and I did read it. I have still not been able to build the courage and commitment needed to read all of Joyce's Finnegan's Wake.

Though I have a BA degree in English, the American Nobel Prize winner, William Faulkner, had never had more than a glance from me in college. However, by the time I was a couple of months into my readings, I had read all of his major works; ***Light in August***; ***Absalom, Absalom***; (a book almost impossible to read, but I read it); ***The Sound and the Fury***; ***As I Lay Dying***; ***Sanctuary***. I read them all, but decided, at the end, that Faulkner's style is a little too personal to him to work for me as a style.

I read popular contemporary novels such as those written by my favorite mystery writers, James Lee Burke, creator of the magnificent, struggling to stay sober, ex-marine and Vietnam vet; occasionally over-whelmed by the horrors of war; a washout from the New Orleans Police Department, Dave Robicheaux, and his big, overweight, violence-prone, pork-pie hatted, pink Cadillac convertible driving, bail-skip-chasing, too often boozed up side kick Cletus Purcell.

Elmore Leonard's funky dialogue also came in for a lot of close scrutiny. I enjoy all the work by Elmore that I read, but his stories pale slightly when held against the rich tapestry of the Louisiana bayou around New Iberia, especially down by the bait shop when owner, ex-New Orleans police officer Dave Robicheaux, asks his old black retainer and boat, fish, and beer wrangler, Batist, to rig a skiff for a customer; all while the bream jump and splash in the warm summer waters near the dock on Bayou Teche, while the heron try to snatch them, and the three-legged raccoon, Tripod, scampers about looking for bits of oyster in discarded shells.

Now that, in my view, is beautiful writing.

***

Many years ago, at a week-long art festival while I was an undergrad at the University of Alaska – Fairbanks, I had the great pleasure of meeting and hanging out with Norman Mailer and Ralph Ellison at a week-long Arts Festival on the campus. This was in the mid-1960s. I had read Mailer's ***Naked and the Dead*** back then, but never Ellison's ***Invisible Man***. Now I have read Ellison's book as part of my research into writing styles for this collection of stories.

Earlier today (November 10, 2019) as I began my final editing on this book, there was an interesting bit of serendipity on NPR about black Supreme Court Justice Clarence Thomas. Apparently, one of Thomas' formative influences was Ellison's ***Invisible Man***. As a reader you should make of that what you will.

These two authors could not have been more different in their personal styles, though their best-selling books were, mostly, within the limits, then, of barely acceptable good tastes in mainstream publishing. During their visit Mailer, a New York City ex-combat infantryman and Jew built in the fireplug shape of a prize-fighter, grabbed the male host at one of the artsy, rowdy parties - held in honor of the two best-selling writers - in a big bear hug. The two men proceeded to dance, hugging, around the floor while both uttered loud growling bear noises.

All the while the black, tall, thin, and reserved Ellison took drags on his holdered cigarette and looked at something in the ceiling above the noisy get-together; ...or so my memory has it.

Like Eudora Welty's 1941 short story, "Why I live at the P.O." these im-ages of contrasting styles have stuck with me for as long as I have thought about the nature of writing and the creation of a robust American literature.

Three of my favorite fiction books in college were ***Catch 22***, by Joseph Heller, ***Cat's Cradle*** and ***Slaughterhouse 5***, both by Kurt Vonnegut. I read them all again. I read a lot of Shakespeare because Juneau, Alaska is a small community full of Shakespeare enthusiasts. There was a festival, and they had all got together to read plays from his "First Canon" when a copy was on loan to our Alaska State Library, Archives, and Museum (the SLAM). I and a friend joined them for a reading of **Titus Andronicus**.

I have become a fan of Seattle writer Neal Stephenson primarily because of his ***Baroque Trilogy***; though his first novel, ***Snow Crash*** is what brought him to my attention. Stephenson's Trilogy is a set of three 1,000 page books describing, in fiction, the roots of the Enlightenment as seen through the eyes of a band of adventurers and trouble-makers. The story unfolds around the time of Elizabeth 1 and her Scottish successor after her death, James 1. In the course of their various "walkabouts," Sir Isaac Newton takes over the minting of English coinage, and much of the City of London burns to the ground.

These adventurers counted themselves as friends of the members of English royalty as well as friends of the members from the lowest ranks of English society. They counted as friends of Sir Isaac Newton. As Stephenson weaves his stories and his many characters among the events and historic incidents of the tumultuous times between the plague and the industrial age, I became very enamored of Stephenson's story-telling practice. ***Cryptonomicon*** is also a favorite.

Among modern American writers the Hemingway style is the easiest to read, but I found myself rejecting the content of ***The Sun Also Rises***, and ***A Farewell to Arms*** as being too refined and a little too sweet for my taste. Then I read ***For Whom the Bell Tolls***,

Hemingway's story of an American volunteer who fought with the liberals and republicans in the Spanish Civil War against Franco's Fascists. I came to the realization that the story is close to being one of the finest works of war fiction that I have ever read. Whatever writing style finally evolves for me, I hope it will carry my fiction across to the reader in as clear and as moving a way as did Hemingway's style in this book.

The movie ***Chinatown***, directed by Roman Polanski, and released by Paramount in 1974 must be noted here because it is the finest example of a movie-maker's story-telling style that I have seen. Guillermo Del Toro's ***Pan's Labyrinth*** comes in as a close second.

Finally, another important part of a writer's style is his or her use of story structure. General considerations of structure are not always obvious in reading because structure is usually relevant only to a particular story. Structure is much more apparent – that is, easier to see - in movies. In my case one movie in particular: ***Cloud Atlas***, - produced by the Wachowskis and released in 2012 - offered me the structural pattern I needed for my collection of stories. ***Cloud Atlas*** linked, loosely, six separate and independent stories over a period of about 300 years. The ***Cloud Atlas*** story structure seemed a perfect model for my attempt to write 10 stories loosely linked over a period of about 100,000 years.

Of the 250 references in my Concordance to my first book ***Cassiopeia's Quest*** and now ***Cassiopeia's Quest - Revelation***, 198 are books, and half of those are works of fiction written in English.

====

The origins of my impulse to write a book-length collection of short historic fiction.

The transformation of these ideas into a book began with a series of large oil on canvas paintings or palimpsests of overlaid images that I began with the help of professional model in 2013. I completed the paintings in time for a personal show that went up on the walls of Juneau's NPR, public radio and multimedia outlet,

KTOO (ktoo.org), in December 2015. These paintings reflect my continuing interest in representing the details of human diversity as individuals struggle within the rigid constraints of culture, tradition and history.

With the completion of this book I hope that I have been able to flesh out my original vision in the telling of these 10 stories.

The stories in Cassiopeia's Quest – Revelation have grown in my mind and imagination over the many decades prior to the time I first applied my number 2 pencil to blank paper and daubed my first brush full of paint on canvas to try and bring each of them, like Prometheus, fully to life.

As I have said many times, I hope all my readers will enjoy this book.

Juneau, Alaska, November 2019

Writer's BIO

Jerry Smetzer lives, writes, paints, and dabbles in politics, business and government. He hangs out with his long time-time gang of friends in Juneau and Fairbanks, Alaska USA. In his private time when not writing, he explores his several and diverse interests in literature, history, visual arts, mathematics, physics and the many achievements of his talented family.

Jerry was born in Kansas and raised on a farm, one among many small farms in north central Ohio. He lives, studies and works in Alaska, residing now in Juneau, earlier in Fairbanks. He has worked jobs in Alaska's bush, biggest cities, and smallest villages – since the age of 16.

He has read deeply and traveled widely all his life, including service as a land survey advisor while a volunteer with the U.S. Peace Corps in Afghanistan. On that job Jerry worked with Afghan surveyors in the field at project sites near rural communities and along trade routes used through history by nomadic tribes on caravans of camels and burros.

In order to maintain their complex and ancient trading relationships these family and clan-based nomads cross the country with the seasons each year. In the early spring their travels take them south from southern Tajikistan in the north across the Hindu Kush Mountains and, finally, to southeastern Iran in the south. When winter snow threatens they will have already traveled through the 12,000 foot passes of the Hindu Kush on their way home.

Jerry has three children born and raised in Fairbanks, now scattered with six grandchildren around North America. He was pleased to have gotten together in Florida with them and with all his extended family for his Mom's 100th birthday in June of 2018.

"Since graduating college, I have wanted to write a full length novel good enough to engage the attention, interests and imaginative capacity of readers to see things far beyond their experience. With the help of all my readers maybe we can make a good start toward my goal, with my published collection of short stories, ***Cassiopeia's Quest – Revelation***.

Photo by Hetty

Made in USA - Kendallville, IN
35862_9781732665644